A. G. STROMBERG

First Class Scientist, Second Class Citizen

Letters from the GULAG and a History of Electroanalysis in the USSR

A. G. STROMBERG

First Class Scientist, Second Class Citizen

Letters from the GULAG and a History of Electroanalysis in the USSR

Richard G. Compton
Oxford University, UK

Alexander S. Kabakaev
Tomsk Polytechnic University, Russia

Michael T. Stawpert
Oxford University, UK

Gregory G. Wildgoose
University of East Anglia, UK

Elza A. Zakharova
Tomsk Polytechnic University, Russia

ICP Imperial College Press

Published by

Imperial College Press
57 Shelton Street
Covent Garden
London WC2H 9HE

Distributed by

World Scientific Publishing Co. Pte. Ltd.
5 Toh Tuck Link, Singapore 596224
USA office: 27 Warren Street, Suite 401-402, Hackensack, NJ 07601
UK office: 57 Shelton Street, Covent Garden, London WC2H 9HE

British Library Cataloguing-in-Publication Data
A catalogue record for this book is available from the British Library.

A G STROMBERG — FIRST CLASS SCIENTIST, SECOND CLASS CITIZEN
Letters from the GULAG and a History of Electroanalysis in the USSR

ISBN-13 978-1-84816-675-2
ISBN-10 1-84816-675-3

Typeset by Stallion Press
Email: enquiries@stallionpress.com

Printed by FuIsland Offset Printing (S) Pte Ltd. Singapore

Foreword

This book tells the story of the life of Armin G. Stromberg, a significant and influential chemist whose scientific work was carried out in the Siberian cities of Sverdlovsk (now Ekaterinbrg) and Tomsk. He was born before the Russian revolution and died after the collapse of the USSR. He experienced the GULAG, contributed to the Soviet atomic bomb project, and founded an academic school of electroanalytical chemistry of high impact, especially within the USSR.

The idea of this book derives from a visit by two of us (RGC and GGW) to Tomsk in September 2006 where we met many of Stromberg's former students, and in particular our future co-authors ASK and EAZ. We were impressed by the tangible legacy that Stromberg had left upon our fellow chemists and friends at Tomsk Polytechnic University, and fascinated to learn that some documents still survived from his internment in the GULAG. The book was realized and flourished because of the support of the Research Centre of St John's College, Oxford and the enthusiasm and interest of its Director, Professor Linda McDowell, which enabled the development of the project to the point where Imperial College Press/World Scientific were persuaded to publish this book. We also thank an anonymous referee of our proposal who asserted that 'you do not need to be a professional historian to write an excellent history book'! We hope (s)he is correct; four of the five authors (but not MJS) are chemists and include the daughter of Armin Stromberg (EAZ).

RGC & GGW
June 2010

Contents

1

The Deportation of the Ethnic Populations of Russia

Russia, despite the name and the popular image, is a nation whose population comprises relatively few ethnic Russians. Historically the Russ, those Scandinavians that migrated into the Novgorod region in the 10th century, are latecomers to the melting pot of peoples that occupy the vast stretches of land between the Ukrainian border and the Pacific rim. Undoubtedly, the political centre of gravity of the Russian state in all the forms it has taken over the centuries was — and remains to this day — the region west of the Urals, which we now associate with modern day Moscow and St. Petersburg.

The gradual expansion of this political power, reached its zenith with the creation of the USSR, the largest state in modern history, stretching out from Russia's European borders in the west, over the Ural Mountains, across Siberia and the Northern Great Steppe, and into the Central Asian regions and beyond to the Pacific Ocean. The expansion of Russian political influence, which led to the subsequent creation of a Soviet Union 'super-state', was often achieved at the expense of the native peoples of these lands. Many of these people underwent the very Russian punishment of mass deportations and, even to this day, many have not yet returned to their native lands. This process was by no means a uniquely Soviet phenomenon, but it was under Joseph Stalin's rule that the number of people that underwent mass deportation reached its high point, and it was carried out with the most ruthless, state-sponsored efficiency.

One ethnic group in particular illustrates the ruthlessness of Soviet deportations: they are the ethnic Germans of Russia. The German-Russians were themselves immigrants into the Russian state, once welcomed and encouraged as settlers in the growing expanse of Russia under the Tsarist regime. Yet they in turn would later become the victims of the same processes that had undone many of the more ancient inhabitants of the Great Plain. In 1941 they, as with many other ethnic groups deemed to be 'non-Russians', were deliberately scattered to the wind and left to do or die in the Siberian hinterland and Kazakhstan.

This book tells the story of one particular German-Russian, Armin Genrikhovich Stromberg. Armin became one of the most influential Russian scientists to work in a field of chemistry called electrochemistry. More precisely, Armin's later scientific work was to become instrumental to the understanding and practical application of a relatively new technique (at the time), that of polarography. His work would help to establish Russian dominance in this field of science for many decades to come. It underpinned many industries such as the metallurgic and even nuclear industries of the Urals, and it even led to the founding of a scientific school in Tomsk, Siberia, from which many famous and influential Russian polarographers and electrochemists would later graduate.

However, before Armin could complete his scientific journey of discovery, he first had to survive the turmoil of World War II, and in particular his internment in an NKVD labour camp, or GULAG. The 74 surviving letters written by Armin to his young wife and infant daughter, Elza, grant us a rare and tantalising glimpse of the daily lives of the ordinary German-Russian inmates of such camps. These were people who, for the most part, considered themselves to be patriotic Russians. Yet, as we shall see, they would still become the target of the state's xenophobia, branded as 'undesirables', and subjected to the Soviet method of punishment, deportation and mass-internment. Their only offence was to have the 'wrong' nationality printed on their documents. But before we begin Armin's story, we should first examine the historical events against which his life was set, and which would eventually bring about calamity for Armin, and many thousands of ordinary people like him.

In 1986, it was reckoned that the USSR contained peoples of no less than 130[1] different nationalities or ethnic groups within its borders. True, this was at a time when Soviet power included the Ukraine, Belarus, the Baltic States of Estonia, Latvia and Lithuania, as well as many of the Central Asian territories that have since become independent sates such as Kazakhstan, Georgia and Uzbekistan to name but a few. Yet, even taking these 'satellite states' into consideration, bringing so many diverse groups of people together under one national banner was no small feat. This figure becomes even more impressive when one considers the number of nationalities and ethnicities excluded from the official list of proscribed entities within the USSR, as well as the number of disparate groups who had either fled the USSR or simply been driven out of existence by the gradual process of Russification. The process of Russian subjugation of the peoples living in the vast expanse of what would later become the USSR took place over several centuries, and there are many examples of resistance to the expanding influence of Tsarist St. Petersburg.

As early as 1571, the Crimean Tartars — Turkish in character, but fiercely independent descendants of the earlier Mongol invasions — sacked Moscow in retaliation for Russian attempts to extend their influence over this region.[2] It was only in 1739 that the lands surrounding the Sea of Azov, a tiny pocket of water hemmed in between modern day Ukraine and Russia, were finally brought under Tsarist-Russian control; and even then this region only existed as an isolated pocket of Tsarist dominion, surrounded on all sides by independent states of varying strength and hostility. When Russian influence began to extend across the great expanses of the Steppes, the nomadic peoples residing there were also gradually forced into submission. For example, the Kalmyks, direct descendants of Genghis Khan and the Golden Horde who still read and wrote in Mongol and followed the Buddhist faith, were finally territorially absorbed into Russia in 1771. As with many of the ethnic groups subsumed by 'Mother Russia', the Kalmyks maintained their cultural

---

[1]*The Soviet Germans: Past and Present*, I. Fleischauer, B. Pinkus, E. R. Frankel (Ed.)., 1986, Hurst, in association with the Marjorie Mayrock Centre for Soviet and East European Research at the Hebrew University, Jerusalem, London, 2.

[2]*The Nation Killers*, R. Conquest, 1972, Sphere.

autonomy for as long as possible, until their semi-nomadic lifestyle enabled the Tsarist forces to occupy their traditional grazing lands and so bring about their subsequent downfall.

The Tsarist strategy for dealing with instances of resistance amongst the non-Russian populations involved a mixture of brutality and persuasion in roughly equal measure. Where there were instances of armed resistance, Russian troops were sent in without hesitation — although with varying degrees of success. For whilst the Crimean Tartars were quickly overwhelmed by Russian troops in the war of 1853, Russian attempts to subdue the Chechens have proved to be an entirely different matter altogether. Despite the six-and-a-half divisions of Russian troops sent to quell the Chechen revolt of 1847–1860, the reduction of the Chechen population to one quarter of its previous size, and the capture of the Chechen rebel leader, Sheikh Shamil (not to be confused with the former modern-day Chechen rebel leader, Shamil Basayev, killed in 2006), the Russian-Chechen situation has still not been satisfactorily resolved to this day. In 1905, the authority of the central Russian government had been severely weakened following the double disasters of the Russo-Japanese War and the events of Bloody Sunday in St. Petersburg. Despite the Chechen rebellion supposedly having come to an end some years earlier, the region chose to exploit the weakness of the prevailing government and declared itself to be an independent state.

Whilst the Chechens were putting up armed resistance, the Crimean Tartars took an altogether different approach for dealing with unwanted Russian attention by simply packing up and leaving. Their common heritage with the Ottoman Empire gave the Crimean Tartars an escape route across the Black Sea, which 231,177 of them took between 1860–1862 in order to flee Tsarist rule. This was an option that Tsar Alexander II was all too happy for them to take. Indeed, the 'voluntary' departure of the Crimean Tartars allowed easy access to Crimean land for Russian settlers and soldiers to occupy, whilst at the same time it solved the problem of what to do with all the non-Russian citizens who had previously held this valuable land that the Tsar desired.

Deportation began to be used as a tool for punishment in Russia in the early 19th century, under Alexander I who, in 1807, deported Jews away from the borders of Russia in the wake of the Napoleonic wars. In this instance, as in later years, these mass-deportations were justified using the pretext of protecting national unity and maintaining military security. The next instance of a punitive deportation of a specific ethnic group occurred during the aftermath of the Crimean War. The conflict between the Crimean Tartars and the Tsars (that had been ongoing even before Russia formerly annexed the Crimea in 1783) finally came to a head with the outbreak of the Crimean War fought between Russian Imperialist forces and an alliance of Britain, France, the Ottoman Empire and the Kingdom of Sardinia. Despite numerous tactical blunders on both sides, the eventual strength of the Russian invasion, the weakness of the Ottoman Empire and the decline in the perceived importance of an independent Crimean state ensured that the war was never likely to end well for the Crimean Tartars. After the war ended in 1856 some 231,000 Crimean Tartars fled the ensuing conditions of persecution and land expropriation by Russian settlers and soldiers between 1860–1862. However, those less fortunate Tartars that were not able to flee and had to remain in the Crimea under Tsarist rule, or who remained in Russian occupied areas during the war, were forcibly deported from their prosperous, fertile and warm homeland into the wastes of Siberia. This process was entirely punitive but it also had the added bonus (for the occupying Russians) of freeing up thousands of acres of rich farmland with which to reward Russian settler-soldiers. This was a model that would be oft repeated throughout the course of modern Russian history.

In 1914, it was the Jews who became the first ethnic minority to be deported in the 20th century. This took place in response to the German victories over Russian forces and the subsequent invasion of the Ukraine and the Baltic states. The Jews were deported out of these areas in order to prevent any possible Jewish collaboration with the invading German forces. However, in 1914 it was not exclusively the Jews who were deported away from their homes in Western Russia for fear of collaboration with the Germans. One other significant group of people that had existed virtually autonomously within Russia for around 150 years previous to 1914 were suddenly uprooted and deported. These were the ethnic Germans of Russia. Up until then, the German-Russians had been a welcome part of Russian society, having been actively encouraged to immigrate into the Russian state by the Tsars themselves. Prior to 1914 these German-Russians had, by and large, prospered under the Tsarist regime. Little did they imagine that subsequent events would see their people enduring a series of catastrophic events that would soon lead to their descent from being revered and protected parts of the state's machinery, to becoming second-class citizens, thrown to the wind with no consideration given to their survival.

Unlike the peoples of the Central Asian parts of the Russian empire, who had to endure conquest and subjugation by the Russian Imperialist expansion eastwards, the Germans of Russia voluntarily emigrated from Germany into Russia. Catherine the Great continued the earlier policies of Peter the Great in attempting to modernize Russia into a western state, and actively encouraged foreigners to immigrate into Russia. At this time the socio-economic value of German immigrants to an aspiring state like Russia should not be underestimated. Prussia had proven itself to be one of the major powers on the continent during the Seven Years' War. With the decline of Sweden in the 17th and early 18th century, Prussia rose to prominence as the military powerhouse of Europe. In July 1763, Catherine issued the first of her decrees inviting foreigners into Russia as members of the state. To entice them further all Germans were granted exemption from military service in the forces of the Tsarina, freedom from some taxes, materials with which to build their homes and were given assistance and encouragement to build a career as farmers. In response to these generous offers some 25,000 Germans entered Russia in the first wave of migration, and over the course of a century more than 200,000 Germans would follow. In return for these concessions from the monarchy, the Germans were expected to use their particularly industrious impulses to help tame the Great Russian wilderness and make profitable the vast expanses of newly acquired territory of the Russian empire.

The success of the German population in acclimatising to a new nation was almost entirely unheralded. In the area around the Black Sea, 60% of all land holdings were in the possession of Germans by the turn of the 19th century, and in the Volga region they formed an entirely new ethnic group. The Volga-Germans were both the largest single concentration of German speakers in Russia and the most fiercely culturally independent. Their success as farmers and landholders produced results that even Catherine the Great could not have dared to hope for. By the late

19th century, Germans on the Volga were limited to owning just four acres per man because their numbers had swelled to such an extent that land was becoming genuinely scarce. However, the communal nature of the farming system that they developed in this region still rendered the land of the Volga-Germans profitable both for the state and for themselves.

Thanks to the various tax and other exemptions that Catherine had imposed to encourage Germans to settle in Russia, the German populations were able to operate with relative independence from the political situation in Russia. In 1874, when their exemption from military service was revoked, a migration out of Russia did occur. Many Germans travelled even further east than their forefathers had done originally or, following the pattern of so many Eastern Europeans at this time, travelled onwards to America for exactly the same reasons that they had originally moved to Russia. However, beside this one, minor exodus out of their adopted homeland, most Germans remained in Russia in splendid isolation of the patterns of Russian politics — that is until the first deportations of 1914 began. Even then, the trials of the Germans living west of the Urals were of little concern to those quietly getting on with their lives on the Volga, in Georgia or in Siberia. That is, until the events that were about to unfold would change the life of every German in the Russian empire irrevocably.

In 1917, perhaps the biggest single seismic shift in Russian history occurred. Whilst the events surrounding 1933 have received a great deal of attention as a year that would go on to shape Europe's future for many decades to come, the bloody civil war of 1917 has left a scar upon the minds of many Russian people even to this day. The consequences of Russian withdrawal from World War I may have had global significance in later years, but undoubtedly the single most significant outcome of 1917 was of course the emergence of the Bolsheviks and the beginnings of their rise to power in Russia. Amongst the German population of Russia there were few who actively supported the Bolshevik movement. The isolation of German-Russians gave them a sense of ambivalence towards the politics of the age. It is possible that their previous experiences of a politically disunited and tumultuous Germany, that had originally led to their departure in the previous century, had grown into a certain 'political caution' passed down through the insular German communities. Either way, the 1917 revolution took its course without significant involvement from the German-Russians — perhaps if they had known the consequences of the Revolution for them they may have acted differently.

At first there seemed to be little hostility between the Germans and the new communist leadership of the country. 1917–1924 was a time of consolidation and the first teetering steps towards building a Soviet Union capable of taking its place on the world stage. Open conflict with a large, powerful internal ethnic group was not one of Lenin's priorities. Additionally, the success of German farmers made them a prime model group to target with the New Economic Policy, enacted by Lenin from 1921 onwards in an attempt to develop a Soviet economy left in tatters after the cessation of hostilities in World War I. It would seem that the same work ethic that had attracted the Tsars to German farmers was to have the same effect upon the new leadership of the nation. Profit through hard work, strong community spirit and a clearly defined sense of identity delineated the Germans from the rest of the Russian agricultural peasantry. However, their social advantages could not protect them forever from either the strong anti-peasant reaction of the communist party, or the fundamental anti-nationalist element of communism. Lenin saw nationalism as a class action, perpetuated by the very classes the revolution was fought against, the forces of reaction. Therefore any minority nationalist group within Russia was seen as a threat to the fledgling Soviet republic and could not be tolerated indefinitely. Lenin's successor would go down as one of the great butchers of history and his attitude towards the ethnic Germans was as intolerant, barbaric and inconsistent as it was towards his own — or any other Russian — people.

The Great Terror that arrived with Stalin's rise to power first affected the German population of Russia in 1935. This was a somewhat later date for their inclusion amongst the long list of Russian people singled out for punishment under Stalin's regime than most other groups. This is, in part, a testament to the German-Russian's reluctance to become involved in Russian politics — they really did consider themselves a separate entity to the majority of the Russian population. For more evidence of this attitude amongst other ethnic groups in Russia one only need read Mikhail Sholokhov's epic masterpiece *And Quiet Flows the Don*, written about the

Don Cossacks but who display the same separatism of many other of the ethnic peoples of Russia. Another factor to consider (although this remains historically uncertain and difficult to prove, thanks to the NKVD's later attempts to erase any formal evidence that might indicate that those 'punished people's' were in any way useful) was the ability of the German population to produce large amounts of vitally needed grain and food stuff for Russia's ailing urban population that was still reeling from Stalin's crash industrialisation. This made the German-Russian farmers too valuable a group to attack immediately. However, their status as not only the very epitome of kulaks[3] but also as potential allies for the new fascist power emerging on Russia's western border, made it inevitable that the Germans of Russia would suffer some sort of punitive action. In 1935 the Soviet army, supported by the NKVD, forcibly deported the German population living within 100km of the border of the Western Ukraine. On the 1st–7th of January that year the men in this area were rounded up into freight carriages and transported east of the Ural Mountains into Siberia. On the 28th the army and NKVD returned and transported the remaining women and children until not a single German was left in this entire region. Their fate in Siberia was of little concern to Stalin or the authorities that arranged for their transportation.

In 1941 Nazi Germany invaded the USSR. Operation Barbarossa, as the German's denoted the invasion, remains the single largest land invasion of all time. The Soviet forces were driven into a rapid and ignominious retreat in the face of the Nazi blitzkrieg. As the Nazis pushed ever deeper into the USSR there was an ever-greater risk of them coming into contact with the German populations of Russia and persuading them to rise against their fellow Russians in mass-defections. The invading forces were not ignorant of the potential of having an indigenous, potentially friendly population within enemy territory. In fact, part of the planning of Operation Barbarossa had anticipated that the 45,000-strong German population of the Crimea would rise up and assist them. Similarly, the Soviet authorities were not blind to the possibility of a German uprising within Russia and the potential damage it could do to the

[3]Prosperous and capitalist peasants who never actually existed as a distinct group, but were merely the most successful members of the peasant classes.

Soviet war effort. Knowing this, the time between 1935 and 1941 was not wasted. Although there is little supporting evidence left behind by the NKVD, there must have been plans made even before the invasion had begun to deport the German-Russians away from Western Russia; only this could account for the speed and completeness with which the Soviet authorities were able to leave almost no German-Russians behind in the areas of Nazi occupation. In September 1941, the following decree was issued by Kalinin, President of the Praesidium of the Supreme Soviet of the USSR:

> *'According to trustworthy information received by the military authorities there are among the German population living in the Volga area thousands and tens of thousands of diversionists and spies who on a signal being given from Germany are to carry out sabotage in the area inhabited by the Germans of the Volga.*
>
> *None of the Germans living in the Volga area has reported to the Soviet authorities the existence of such a large number of diversionists and spies among the Volga Germans; consequently the German population of the Volga conceals enemies of the Soviet people and of Soviet authority in its midst.*
>
> *In case of diversionist acts being carried out at a signal from Germany by German diversionists and spies in the Volga-German Republic or in the adjacent areas and bloodshed taking place, the Soviet Government will be obliged, according to the laws in force during the war period, to take punitive measures against the whole of the German population of the Volga.'*

The 'whole German population of the Volga' was around 64% of the total population of the region, some 576,000 people in 1933. The Volga-Germans are simply the most noticeable German population in Russia to be targeted. They had become an ASSR, Autonomous Soviet Socialist Republic within the USSR, and had continuously shown support for the Soviet regime in elections: 99.7% of the Volga-German population voted for official candidates in the 1938 election. Still, none of this prevented the Volga-Germans from being deported across the interior of the USSR with many of them ending up in Kazakhstan and the Central Asian steppe.

62,000 square miles of land was left vacant by the removal of the Volga-Germans, an area 17 times the size of Cyprus or Czechoslovakia and Albania combined; yet these people account for only one small part of the estimated 1,650,000 people deported for being of the incorrect nationality during World War II.

A difficulty for historians wishing to study these events is the lack of reliable evidence left by the NKVD, the Nazis occupying forces or the people of Russia themselves. The 1926 census is one of our most accurate sources of data with which to obtain figures relating to the population of Russia and her surrounding regions, as it includes information about the areas in which people were living in Russia, unlike the subsequent 1931 census. The best estimates available place the number of ethnic Germans living in Russia at 1.25 million, 1.08 million of who lived in the European areas of Russia, which were eventually overrun by the Wermacht. The estimates for the numbers of Germans deported in 1941 ahead of the Nazi advance are even more vague. The best estimates vary between 600,000 to 900,000 Germans deported beyond the Urals in an attempt to prevent them bolstering the Nazi German army. In addition to the lack of documentary evidence left by the NKVD concerning this matter, any German-Russians unfortunate enough to have been overrun by Nazi forces were subsequently conscripted into service for the Wermacht. Therefore, they were obviously reluctant to talk about this afterwards — assuming that they survived the experience. For they faced death not just at the hands of the advancing Nazi forces, but also at the hands of the Russian Red army, who often sought retribution against their own people — suspected of collaboration — in occupied areas (possibly best represented in the film *Come and See*, which spares no one in its portrayal of the Russian partisans' advance across the Ukraine). Added to this confusion, a little known consequence of Operation Barbarossa was Stalin's flight out of Moscow to a retreat further east of the German advance. So what little direct authority that was wielded by the Kremlin was completely nullified just when it was most needed.

A point to consider, not as an excuse for the acts of Soviet authorities but as a way of contextualising their actions, is that in Britain and America people of German or Japanese descent were similarly interred. Over 100,000 Japanese-Americans were removed from their homes, stripped of all but their most precious possessions, and were dispossessed of their property in the aftermath of the Pearl Harbour attack in December 1941. Of course the difference between the imprisonment of British and American citizens who were nationals of hostile nations is that the conditions of their transportation were far more humane and sanitary than the brutalisation suffered by Soviet citizens. In the Soviet evacuations of 1941, 400 children are reported to have died in one freight carriage alone and whilst the possibility of eventual release from exile was open to British or American citizens, the German-Russian exiles were not given amnesty for their alleged 'crimes' until after 1955 and were only absolved of the stain of collaboration and treachery from 1964 onwards. The barbarity that could be shown towards the ethnic peoples of Russia (some of whom counted themselves amongst the most loyal supporters of the Soviet regime, others of whom were totally indifferent towards the new powers) was characteristic of the breakdown in civilisation that can be witnessed during the 20th century. However, it is important to place these events in a broader context; we are not dealing with a holocaust. The peoples of Russia who were forced into deportation, deplorable as this is, were not sent into Siberia for extermination. True, Stalin cared very little if thousands did die of starvation, exposure and the brutality of some of their guards, but first and foremost the camps were designed for isolation and for economic output. In many instances being sent beyond the reach of the Nazi advance saved the lives of the deportees who might either have been killed in the initial fighting, forced into German prisoner camps — and the treatment for the 'sub-human' Russians was far worse than for British or American prisoners — or made to fight for the Nazi German army. Then, assuming that they survived the horrors of the Eastern Front and 'liberation' by the Red Army, these unfortunates could expect to suffer the humiliation of being executed as a traitor to their own people. Faced with these options perhaps a lengthy exile in Siberia did not seem the worst of all possible outcomes.

In fact, the German peoples of Russia seemed to have made the best of a bad situation. There is a Russian saying: *Немец как прут ивы — посадите его хоть где, и он вырастет*: *'A German is like a willow tree — plant it anywhere and it will take'*. In the face of the harshness of Siberian life, the pioneering spirit that had previously enabled them to thrive in the Russian wilderness emerged again. It was the 900,000 deported Germans who made the most of what the century threw at them, and in many places have leant a distinctly European character to many Siberian cities and regions, far beyond the Europe-Asia border in the Ural Mountains. This is the story of one such man, a German-Russian who adapted to his new world and left one of the most important series of letters, detailing daily life in a NKVD GULAG to remain in existence. This is Armin Stromberg's story.

2

'A Willow in Siberia': The Family History and Early Life of Armin G. Stromberg Pre-1937

Introduction

The concepts that placed the highest value on a human life, along with ideals such as an individual's right to freedom and equality that we take for granted in modern Europe, are the result of a long and sometimes difficult journey. Often during the development of a 'free society' in one country or another there would appear an opportunity for peace and stability, only for that country to pass into revolution or class war, in an attempt to find simple solutions to complex problems.

Russia at the beginning of the 20th century was no exception. World War I emasculated villages and brought ruin and grief to all corners of the immense and unsettled Empire. The majority of Russian citizens were largely illiterate — an excellent ground in which to sow the seeds for the revolution of 1917. The consequences of the 1917 Revolution were years of civil war, repression and state terror, lasting more than seven decades. The casualties of the cataclysmic events brought about by the initial revolution were substantial, as is often the case during such sudden and violent social and political upheavals. The machinery of dictatorship, facilitated by the much-feared secret police, made the lives of a great many people from every walk of life very dangerous indeed during the development of the fledgling super-state. In the 1930s, under the watchful eye of Joseph Stalin, the lives of tens of millions of Russians were crippled, and 'disappearances' or NKVD killings commonplace. Those ordinary people not arrested, killed or 'disappeared' lived with the ever-present fear and understanding that they too could be arrested at any moment on the slightest pretext.

Many families were oppressed by fear and had to hide information about their relatives, their friends or even their long-dead ancestors from beyond the borders of the USSR. They had to destroy correspondence, family archives, photographs, in fact anything that could somehow prove that they shared any connection with the 'enemy' classes. If such 'incriminating' items were found during a search of their homes people were often falsely charged with espionage or treason — with dire and often fatal consequences.

The Stromberg family — German-Russians of Estonian origin — were no exception. Expecting to be arrested at any moment in the late 1930s, the family destroyed all diaries and letters written in the German language, their photographs were cleansed of any German inscriptions and the most valuable of these German documents were hidden behind other portraits. The pre-revolutionary history of the Stromberg family has remained hidden throughout the long decades of the Soviet regime. Now that those fearful times have gone we have the chance to restore these pieces of the past that sketch a tragic picture of the lives of people devoted to their fatherland and imbued with a sense of kindness and responsibility to the surrounding people of their adopted homeland. Armin Stromberg was descended on both his father's and his mother's sides from large German-Russian families. His ancestors were of Baltic-German origin, as distinct from the Volga-Germans mentioned in the previous chapter and who have been the subjects of more widespread historical study. It is therefore necessary to make a small digression to look at how the Baltic-German populations in Russia came about and their similarities and differences from their Volga-German cousins.

Ancestry and Nationality

The first significant increase in the German population in Russia happened at the end of the Great Northern War with Sweden (1700–1721). As a result of the war, Tsar Peter the Great (1672–1725) added territories containing large German populations to Russia, particularly Estland (now northern Estonia) and Livonia (now southern Estonia and northern Latvia). It seems probable that it was during this period that the Stromberg family's ancestors became Russian citizens. Later the Empress Catherine II (1729–1796), who was keen to modernize her empire along the European model, sought to attract tens of thousands of European farmers with the decree of 1764, which organized large colonies in the Volga region ('between Saratov and Astrakhan', as the decree states) from whence the Volga-Germans later arose. A few years after that, following a further decree in 1766, there was another influx of Germans into the Russian occupied Baltic States from which the Baltic-Germans of Russia are descended. In 1772 the Polish-Lithuanian Commonwealth was divided up between Austria, Prussia and Russia. The next increase of the territory happened following the victory over Napoleon's troops in the years from 1812 to 1814, when the entire Kingdom of Poland became an annexe of Russia. Thus, by the middle of the 19th century the Russian Empire had expanded to include extensive parts of Poland, a large part of Livonia and Estland (Estonia).

The German population in these territories was significant. For example, Baltic-Germans comprised about half of the urban population of Estonia at that time. The literacy level in these Baltic regions was high with more than a 90% literacy rate in the Estland and Livland provinces, compared to an average literacy rate of only 15–20% in the rest of Russia, rising to only 50% in the largest Russian cities. Since the times of Peter the Great, the social, political and economic influence of the German-Russians became increasingly pronounced. After the 1721 Treaty of Nystad, Tsar Peter the Great returned all estates to their former owners in the newly conquered Baltic territories. He also restored many privileges back to those members of the Baltic nobility and merchants willing to swear an oath of loyalty to the Tsar. Since then the Baltic-Germans have served Russia. Unlike the predominantly agricultural Volga-Germans, the Baltic-Germans came to occupy increasing numbers of posts in government, and by the mid-1800s some had even risen to serve in ministerial positions. Baltic-Germans also made up a significant part of the Russian scientific community as they had both the opportunity to train at leading

European universities, and were also able to read the majority of the scientific literature that, at the time, was predominantly written in their native German language.

After the outbreak of hostilities between Russia and Germany during World War I, hostility and suspicion towards German-Russians became widespread amongst the indigenous Russian population, but the focus of this xenophobia fell mainly on the Volga-Germans due to their more concentrated population density and socio-economic dominance of the large Volga area in the southern farming heartlands of Russia. Unlike the Baltic-Germans the Volga-Germans were politically inactive bordering on indifferent; however, as we saw in the previous chapter, this political neutrality would not save them. For example, whilst both the Baltic- and the Volga-Germans were called to serve at the front in World War I, the Volga-Germans were often sent to do hard, manual labour alongside Turkish prisoners of war, and upon returning back home they were treated almost as if they were the enemy. The more urbane Baltic-Germans seemed to be treated somewhat better.

Both of these groups were subject to anti-German legislation, for example German newspapers, public organizations and schools were closed and, by the decree of Tsar Nicholas the II (and last Tsar of Russia) it was prohibited to use German in public places. However, after the Bolshevik revolution of 1917 it was the Volga-Germans who bore the brunt of the resulting problems as members of the prosperous farming peasantry. The surplus appropriation system (*продразверстка*), and the nationalization of all organizations resulted in the famine of 1921, and seriously affected both the people and the economy of the farming Volga-region. When the Volga-German Autonomous Soviet Socialist Republic was disbanded 400,000 Volga-Germans were deported in 1941 to the Urals, Siberia and Kazakhstan. These are the people that Armin Stromberg describes as ‘German peasants’ in his later letters from the GULAG.

Armin Stromberg’s parents, Heinrich (02.07.1881–27.09.1914) and Magda (18.06.1885–12.08.1972) both came from large families of Baltic-Germans and both received their higher education in St. Petersburg. This was both their families’ tradition and was also somewhat *expected* since the Estonian-Germans were mainly members of the gentry or merchant classes.

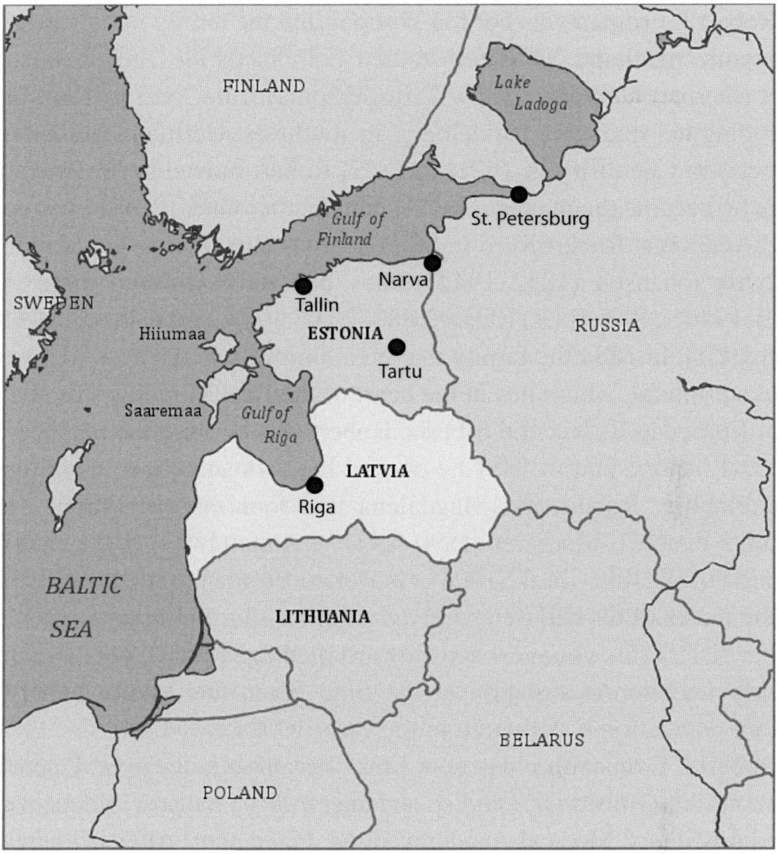

A map of the Baltic region showing the relevant places mentioned in the text.

It is worth mentioning that Armin was a member of the third generation of chemists in his family. One of the earliest of Armin's ancestors that we know of was his great-grandfather Carl Johansen-Klabass (07.06.1818–24.10.1893) and it seems that he had dedicated his life to science, working as a doctor in St. Petersburg in the Botkinskaia and Petropavlovsk hospitals. Armin's grandfather (Magda's father) Robert I. Erdman (07.06.1844–21.02.1909), devoted his entire life to chemistry. As was only recently discovered from a file of Heinrich (or 'Henry' as he is sometimes referred to) Stromberg in the Russian Military Archive, Robert Erdman was also a member of the Estonian gentry.

Robert I. Erdman was born in Wolmar into the family of a professor of therapeutic medicine. He later studied both medicine and chemistry at Derpt (Dorpat) University (now Tartu, Estonia) from 1863 to 1868 before continuing his studies in Heidelberg, in southwest Germany. He defended his thesis in Chemistry in 1871. In 1872, Robert moved to St. Petersburg, where he became the manager of a candle factory until 1884. It was during the 12 years that Robert lived in St. Petersburg that he married Magdalena Karlovna Johanson (1852–1942). They had three children there: Ernst (1879–1940?), Robert (Jr) (02.09.1880–25.03.1908) and Ella (03.11.1882–29.08.1921). In 1884 the family moved a long way to the east, to the town of Yekaterinburg, which lies in the heart of the Ural Mountains on the border of European Russia and Siberia. Robert (Sr) became the manager of a chemical factory, and in 1887 he opened his own analytical laboratory. In Yekaterinburg, Robert and Magdalena had four more children: Magda (Armin's mother), Margaret (1890–1937?), Anna (1893–18.03.1940) and Edward (14.03.1895–26.02.1904). As Robert Erdman (Sr) was an Estonian nobleman, six of his children received a higher education including Magda and her sisters (his youngest son Edward died aged nine). This was a particularly uncommon situation at the time given that barely half of the Russian population was literate in the early 1900s.

Robert I. Erdmann's eldest son, Ernst, became a judge in St. Petersburg; his second son, Robert (Jr), studied and later took up the post of docent at the Imperial Military Medical Academy in St. Petersburg. All four daughters, Ella, Magda, Margaret and Anna, graduated from the Female Bestuscheff Courses in St. Petersburg, which was a university specifically for women. Robert Erdman Jr. was the person that acquainted his sister, Magda, with her future husband. Wise beyond his years from his experiences in the Russo-Japanese War, the 27 year-old staff physician of the Military Medical Academy, Heinrich H. Stromberg, was Robert's friend. This is how he and Magda were introduced and married in 1908. Sadly, in that same year Magda's brother Robert committed suicide over a case of unrequited love.

Armin's Parents Heinrich and Magda

Heinrich and Magda's story is fleeting, interesting and deserves a separate mention. Heinrich Samuel Wilhelm Stromberg was born on the 02.07.1881 in Hungerburg,[1] a small Estonian town on the northern coast where the Narva River flows into the Baltic Sea. His father, Heinrich Carl Stromberg, was a merchant, but died of consumption a mere four years after his son's birth. Later, the younger Heinrich's son, Armin, would also lose his father at the age of four.

Despite losing her husband, Heinrich's mother, Jilie Marie, managed to raise four children. Heinrich graduated from the Narva gymnasium in 1899 with a silver medal. Besides the usual subjects of mathematics, physics, history, Russian and geography, Heinrich also studied Latin, German, French and Greek at that school. The reason why he obtained a

Heinrich G. Stromberg, Armin's father.

[1]Hungerburg was the German name. The town is now known by the Estonian name, Narva-Jõesuu.

silver medal instead of a gold one was that he only achieved a four out of five mark in geography — although he would go on to study this subject in a very different way in the future, serving his country in three wars everywhere from China's Manchuria, through Uzbekistan in Central Asia, Georgia in the Caucasus and finally in the Balkans.

However, before embarking on his military adventures the young school-leaver entered the St. Petersburg Military Medical Academy, where he learned surgery. For the first two years there, Heinrich earned his living by giving private tuition and translating and sketching tables for lectures. In his third year of study he was awarded a state scholarship.

Heinrich's fourth year of his medical training was spent in the Abas-Tuman military hospital located in a small town in hills of Georgia, and from there he was sent to an infirmary in New Margelan city.[2] On the 29th of February 1904, whilst only a fifth year medical student, Heinrich volunteered to go to Manchuria, to the seat of the Russo-Japanese War, at the request of the Red Cross together with the 3rd St. Petersburg' Flying Hospital. He arrived in Harbin on the 18th of March having journeyed for more than two weeks by train, and spent one and a half years at war. Together with his flying hospital, he took part in all the major battles of that time. He was awarded the Cross of St. George (silver, fourth class) after battles near Liaoyang, and another silver medal after a battle near Mudken.[3]

Upon his return to St. Petersburg, Heinrich worked in several different hospitals and in the autumn of 1906 he passed the examination to become a 'physician' *cum eximia laude*. As Heinrich had received a state scholarship in 1902 he was required to serve in the army, but instead the Military Medical Academy decided to direct him towards further study. Between 1906 and 1909 Heinrich worked as an intern in the hospital's surgical clinic under Professor S. P. Fedorov. By 1908 Henry had successfully passed the Doctor of Medicine examination and was elected as a full member of the Pirogov' Russian Surgical Society.

It was around this time that Heinrich was introduced to his friend Robert Erdman's sister, Magda. Heinrich and Magda's marriage proved to be a happy one despite the fact that she was not the usual woman of choice

[2]Renamed in 1910 as Fergana, Uzbekistan.

[3]The Manchu name of what is now Shenyang city.

The proud parents, Magda and Heinrich Stromberg, holding a young Armin.

for a man born of a German-Russian merchant family. For one thing, Magda was a member of the nobility. Furthermore, she had obtained a university diploma in both mathematics and physics. Thus Magda did not fit the principle of the 'Three Ks' — Küche, Kinder, Kirche (kitchen, children, church) — which was thought to be the best practice by which to choose a wife in that time. Apparently this resulted in rather frosty relations with Henry's parents, but it never affected their life together. Their short family life was apparently full of love and happiness, as Magda wrote in the album of her daughter Elga:

> *'This summer of 1914 Heinrich daily took you to into his arms and admired you with such tenderness. He found you pretty and stated that "such beautiful children as ours is proof of the fact that the parents form a good pair".'*

On the 7th of November 1909 Heinrich defended his thesis, and soon afterwards he left to undergo further practical training in various clinics in Germany for two years, 'by royal permission' dated 30th of January 1910. He left together with Magda on the 10th of February and it was during this trip that Armin G. Stromberg was born in Breslau,[4] then one of the large German cities. With the benefit of hindsight we can certainly see now that this place of his birth promised nothing good for the future of the boy, but at that time who could have imagined the drastic changes that were about to happen in the next decade?

By the end of December 1911 the trip was over and the family returned home to St. Petersburg. Heinrich was ordered back to the surgical clinics, and it seemed that a happy family life would ensue. However, in 1912 the Balkan Wars against the Ottoman Empire began, and Heinrich left as a doctor in the army of the Red Cross, having lived in St. Petersburg for only a short amount of time from January to October of 1912. In the summer of 1913, after returning home from the second war in his lifetime, Heinrich's career progressed rapidly and he became the assistant professor for clinical surgery at the Military Medical Academy. In the spring of 1914 his daughter Elga (04.05.1914–03.04.1962) was born, and the entire family spent the summer at their dacha in the suburbs of St. Petersburg. This was their last happy summer together.

On the 1st of August 1914, Germany declared war on Russia and in so doing quickly turned the lives of the entire country upside down. On the first day of the war Heinrich left to assemble a field hospital. He went to the front, and he and his team's work saved many lives every day, but on the 29th of November 1914 Magda received a telegram:

> *'Doctor Stromberg was killed twenty seventh September. New-Alexandria.[5] Body sent to Lyublin. Railroad car 325. Call manager of auxiliary unit of the fourth army. The staff of grenadier corps. Doctor Berezantsev.'*

[4]Now named Wroclaw in Poland.

[5]Now known as the Polish town of Pulawy in the eastern Lublin province.

Later, two of Heinrich's colleagues wrote these letters to Magda, explaining events in more detail, and another colleague published his obituary in the Proceedings of the Surgical Clinics of Professor S. P. Federov. Their full text is laid out below:

Letter of Doctor Berezantsev to the Widow Magda R. Stromberg

Doctor B.A. Berezantsev
Acting army,
The staff of the grenadier corps
1914 Oct. 29

Dear Madam!

In response to your telegrams and the letter that I received today, I can report the following: Your husband was walking along the bridge across the Vistula River near the city of New-Alexandria on September 27th at approximately 6 p.m., as the battle raged on the left shore of the Vistula, and our troops were retreating in panic. At this time the bridge was under fire by German heavy artillery. One of the projectiles fell onto the bridge near to where your husband was walking. Both of his feet were badly injured by flying shrapnel. He was immediately placed on a stretcher and sent to the dressing station, but on the way there he passed away due to the heavy loss of blood.

Given that the enemy was successfully advancing and was bombarding the city with heavy artillery, I hurriedly evacuated all of the injured from the city, without forgetting those who had been killed as well.

I put your husband's body into the separate railroad car 325559 R.O., and it was sent to Lublin. His belongings, weapon and money were accepted in deposit by Vasilyev, a medical assistant of 2nd grenadier artillery brigade in the presence of the corps doctor Grenaderskiy.

After sending the body of your husband I notified you and the manager of the medical service of the 4th army by telegram, and *requested to keep the body until your arrival in Lublin. I wasn't able to trace the route of the body of your husband, since I was responsible for evacuating all of the injured (there were about 4000 of them) from New-Alexandria city, which was constantly under fire by heavy artillery, and it was necessary to send the trains in haste.*

On September the 28th the corps left to the north, and rumours reached me that the body of your husband was delivered to Kholm city to the military infirmary. I do not know if the body arrived in Yurev.

I would like to express my sincere condolence for the grief you have suffered, I can tell you that your husband was an example for all doctors of the grenadier corps and the Russian army: he worked coolly under terrible fire in combat from August 11th to 12th near the Goraets, remained in captivity with those who had been injured and returned from this capture to continue his work — to serve the suffering soldiers.

Furthermore he perished at his medical post. He may have left forever, but our memories of him live on in us, and the fatherland will not forget him!

My sincere respects to you and always ready to help.

Dr. Berezantsev

A Letter about Heinrich H. Stromberg's Death, Written by Vladimir Mamontov

26/10/1914

Dear Magda,

Almost a month has passed since Heinrich's death, but I didn't think it would be a good idea to write to you any earlier. I was worried that it would only make your suffering worse if I sent a letter immediately after that telegram which contained the awful news.

On the morning of September the 27th your husband and I crossed the Vistula River near to the city of New Alexandria. A dressing station was set up in the village of Góra Puławska and work started immediately under the leadership of Heinrich, who also took an active part in the proceedings there. Towards the evening it became clear that our corps would have to retreat. Then Heinrich, calm as always, led all of the medical units across the Vistula River and into the city of New Alexandria in order to save the injured and the medical staff. But, since the battle was still raging, your husband left the train of carts with a young doctor, and he, along with two medical orderlies took some bandages and set off once more for the other bank of the Vistula River. It was at that moment that Heinrich and I parted. Everything else I only know from what I have been told. Half an hour to an hour later, not longer, as Heinrich was crossing the bridge on his way back to New Alexandria, one of three bombs fired by the German heavy artillery hit him and ended the life of one of the finest men I have ever met.

That's all that I can tell you about this sad day, the 27th September. It's not just sad for his friends and relatives, but also for all the people who had only a fleeting contact with your husband. We value the fact that he was a talented doctor, a man of duty who was sympathetic towards human suffering and always ready to help however he could.

According to the orders of the Commander in Chief you have the right to have Heinrich's body transported to your permanent home. For this you will need to send a statement of agreement to the Headquarters. In it you should specify the date of your husband's death, where he served and where he was buried. Your husband's body was taken to Lublin and buried there. The transportation of the body will be at the government's expense.

Please accept my deepest sympathy for your grief; I'm always at your service.

Yours sincerely,
Vladimir Mamontov

The following was written in 1915 by Heinrich's colleague Dr. Guenter and was published in the Proceedings of the Petersburg Surgical Clinics:

> *Heinrich H. Stromberg was born in Narva in 1881. When he was four years old, his father died of consumption. His mother was responsible for the children's upbringing, which was especially difficult due to a lack of money. In spite of this, she succeeded in giving a good education to all her children, surrounding them with her tender love in their own home, and most importantly, she taught them the concept about debt to the motherland and about striving to do good. Thanks to his mother, the poverty in which Heinrich grew up had no negative effect on him, and he had fond memories of his childhood. When he was in the final years of high school, he was already aware of the family's difficult situation, and he decided to help his mother, and earned money by giving private lessons. This did not prevent him from graduating from the school with a medal, and he was one of the first to do so.*
>
> *In the autumn of 1899 he joined the Imperial Military Medical Academy. In his first years Heinrich earned money for his living costs by himself, continuing to give lessons, doing translations, and sketching tables for lecturers. His scientific studies did not suffer from this. He knew how to manage his time well, and how to fulfil his numerous responsibilities without overstraining or hurrying himself. He was also a really cheerful youth, a loyal and dear friend, and good company. In the third year he was awarded a government grant, which supported him financially, and simultaneously gave him the possibility to devote his time completely to the science he loved. Even then, his disposition towards surgery was revealed when he started clinical classes, and decided to dedicate his life to surgery. He fell in love with his future speciality early, but it would be a mistake to claim that he became a specialist during his student years. However, he wasn't just interested in clinical work, but also in purely scientific research in the fields of pathological anatomy, general pathology and bacteriology. His main aim during his student years was to get the best and widest general medical education possible. But even that wasn't enough for him. He willingly spent his free time*

reading in the fields of philosophy, literature, history and arts without committing himself to any particular theory, and as a result, attained a wide education and independent views characteristic of a well-educated person. During the summer months he asked to be sent on placements to the most distant outskirts of our native land: to the New Margelan city [now Fergana in Uzbekistan] and to the Caucasus, where he developed an interest in distant countries and a love for nature. Heinrich's graduation from the Academy coincided with the beginning of the Japanese War. It is no wonder that Heinrich, with his lively energetic nature, could not stand his scientific work, or the preparation for his state examinations, when blood was flowing in the distant parts, and young forces were needed. A job opportunity opened up before him in the advance detachment of the Red Cross, the real work of supplying first aid to the injured under hostile fire, the worthy vocation of doctor in the field of conflict, where he was the single creative bearer of the principles of love and aid amongst utter destruction and bitterness. It is no wonder that he took this path; only a few carried out the debt of a doctor in wartime as piously as he did. Heinrich participated in a large battle near Liaoyang and Mudken [the Manchu name for Shenyang City], where he earned the Georgievskiy Cross; he always worked in the foremost positions and took the burdens of this heavy war.

After returning to Petrograd, Heinrich became a volunteer in the German Aleksandrovsk and Evangelical Hospital, and then started his final state examinations. Heinrich won a competition for the post of physician in the Academy and entered the Hospital Surgical Clinic of Professor S. P. Fedorov.

He then decided to start his own family, and married his best friend's sister, Magda R. Erdmann, who shared his enthusiasm and interests in everything, and she was his best friend for the next six years of their matrimony.

As can be seen from the above, Heinrich was a relatively experienced doctor when he entered the clinic, as he worked both during the war and in the hospital. Now new horizons of possibilities were opened for him: to combine the practical work of a surgeon near the bed of a patient or in the operating room with the purely scientific researches of the laboratory and the anatomical theatre. This desire was a thread of all his further works; it also influenced his thesis, which was concerned with fundamental clinico-anatomical research. Heinrich proved to be a bright and talented representative of the scientific direction of Russian surgical science, which was entrusted by the great teacher Pirogov. Heinrich also worked on the same themes during his foreign missions, where he studied bacteriology, the clotting of blood, and roentgenology in different clinics, and became acquainted both with the scientific and the everyday life. Heinrich did not become a blind worshipper of everything foreign, he preserved his independence, clearly recognized all negative moments, such as the insufficiently civilized treatment of patients, who were considered only to be scientific material, and they were subjected to painful studies that were sometimes unnecessary for their treatment. The understanding that the interests of patients' health should be of the utmost importance for the doctor always inspired his activities. As a man of deeds rather than words, he rarely spoke his highly humane views, he did not say grandiloquent phrases, but everything morally low outraged him, and nothing would force him to participate in it. Thus, for instance, he stopped studying in one of the clinics, where the treatment of patients differed fundamentally from his views, and he did not feel shy to report the reason for his withdrawal to the professor. On the other hand, the seething scientific life of the West had a greatly beneficial effect on him, broadened his horizons, and trained him for even more systematic and sequential thinking. Again, his leisure hours were not hours of idleness for him, as he thoroughly acquainted himself with many aspects of West European life and with the remarkable monuments of arts, with the beauties of nature.

On returning to Petrograd he again returned to his previous studies, and in his free time took delight in the happiness and comfort of family life. However, this happiness did not last for long. The war began in the Balkans, and Heinrich went there, where the hand of surgeon is required the most. He entered the Kaufman Community detachment of the Red Cross as a doctor and departed with it into Bulgaria. Here he participated, among others, in the assault of Adrianòpolis [now Edirne]. His communications 'About the activities of the military field medical forces of the Russian society of the Red Cross in the Balkan theatre of war', 'About the action of a pointed bullet according to data from test firing and according to observations during the assault of Adrianòpolis' and 'The organization of operational aid in the line of main dressing stations' became widely known in medical circles.

Heinrich returned to the clinic having gained experience from the second war, but he stayed there for just over a year, and he obtained the title of assistant professor in this time. Young doctors worked under his management in winter, then in summer he managed clinical surgical barracks. A further scientific career seemed to be assured, especially given that he was nominated to the post of the assistant of the clinic by Professor S. P. Fedorov.

But the great world war began, which became fateful for him. Heinrich left on the first day of the declaration of mobilization, where his debt to his native land called him. But what we do know about his last months gives us the same impression as before, that he was a selfless, courageous worker, and placed his debt above everything else. Whilst the hospital he managed, which was located in the Radognitsky convent, was under attack, was bombarded by a shower of bullets and shrapnel, Heinrich personally erected the flag of the Red Cross on the dome of the cathedral; he approached the enemy, and asked them to spare the nunnery. When the convent temporarily fell into the hands of the Germans, Heinrich managed to establish the correct relationship between the enemy and the staff of the Russian hospital, and managed to maintain its independence. Up to 300 injured people were admitted a day, who were aided by Heinrich and his assistants, who in doing so put his life in constant danger. He stepped on the bridge as the river Vistula was under fire, and as an exemplar of bravery, said to his subordinates, 'Let's show that we're not cowards'. On seeing how the fire was intensified, Heinrich understood the danger and sent his medical orderlies back, and he himself went further, but was hit by a shell, which crushed both of his legs. Two orderlies took and carried him to the first aid point. When they put him on the stretcher, he quietly said: 'I can't do anything';

after that he could only moan weakly, but soon fell silent; he died of a hemorrhage as he was being carried. That's how his life ended at the worthy post. His life was short, and he did not have time to archive everything that was expected of him, but even his short period as a doctor made a big difference and saved numerous lives, and spared many tears. But any future studies of the pathology of retroperitoneal tissue, the coagulability of blood, the surgery of pulmonary diseases, the injuries of modern weapons and the organization of medical aid in wartime can hardly ignore his works.

However, his achievements were not only limited to this. Heinrich was a military surgeon with great experience, a doctor who was passionately devoted to his work, and who loved his field of science and profession. He clearly was a well-rounded person with strong convictions, a person of perfect morality, who did not accept compromises or base motives. He was a man of few words who boldly and confidently followed his chosen path, without being influenced by others, and without being distracted along the way.

Heinrich was also a loyal and sympathetic friend, a tenderly loving father and husband and a grateful son. It seemed as if fate were smiling on him, and as if domestic bliss and a promising career were awaiting him; it seemed as if he could make much more of his career, and in doing so, contribute to national science. A stray shell destroyed all our hopes.

His body was brought to Yurev (the Russain name for what is now Tartu, Estonia), where his widow lived with their two small children; and he was buried with the military honours. Rest in peace, our unforgettable friend and comrade.

Dr. Guenter.

A World Turned Upside Down

From the beginning of World War I, Magda's world began to fall apart, along with the lives of other people in Europe and Russia. If we think of her in light of the new realities of World War I Russia we can only imagine the problems that she faced.

Born in Yekaterinburg to a wealthy family of the gentry, Magda R. Stromberg (née Erdmann) had never had to work for a single day in her young life. Yet, up until 1914, she had lived throughout her life in relatively good conditions in terms of both her funds and her relations with the government and the people of Russia. However, soon after the outbreak of World War I this was no longer true. The war with the German Empire caused all German-Russians to be considered as second-class citizens. The use of the German language was outlawed in public. The reward for hundreds of years of loyal German service to the Russian state was about to become a living hell.

With Heinrich dead, Magda was left alone, with two children, a housemaid and a nanny. They had never owned their own house or flat, but had rented different places in St. Petersburg. Obviously she could not afford to do this anymore. Luckily, not long before Heinrich's death, Magda had moved to Tartu, where some relatives lived. This large Estonian city was full of Germans who suddenly had to speak Russian even to simply buy a loaf of bread. So Magda started to teach Russian to them, and was thus able to earn a little extra money.

In the time it took Magda to arrange for her state pension from her husband's death the 'state' was still in existence — but only just. On the 27th of May 1915 the family moved to Võsu, a small village in Estonia with a beautiful beach overlooking the Gulf of Finland, some 36 miles northwest of Rakvere. They spent the summer of 1915 there, but closer to the autumn of 1916, it became apparent that the situation in Petersburg was deteriorating rapidly.

Heinrich and Magda had rented a flat in St. Petersburg that contained some of their expensive belongings. In order to rescue these items Magda, together with her children (Armin and Elga), a housemaid (an Estonian woman called Minna Helman) and Elga's nanny, Liza Kelman (*'a calm pretty Estonian from Derpt'*) left Võsu. They spent a few days in St. Petersburg, arranging the 'safe' storage of the expensive furniture. Magda even pawned some of her belongings, and just packed a few suitcases as she was only planning to leave St. Petersburg for a short period *'in order to wait for the chaos to pass'*.

From St. Petersburg Magda and her charges went to stay with her sister, Ella Kler, in Yekaterinburg — and managed this just in time before subsequent events brought about the total collapse of the state. They had planned to return to St. Petersburg, but the situation there grew steadily worse. It is unclear how Magda managed to make the trip to Yekaterinburg, with the railways being in a state of complete disorder, but soon after arriving in Yekaterinburg Magda realized that despite losing the flat and all their belongings the move had most probably saved all of their lives.

Revolution and a Second Shock for Magda

With the Revolution a new era of Russian history began, and one that for German-Russians such as Magda and her family would be worse than anything that went before. Since the beginning of World War I, people of German nationality or ancestry had been treated as second-class citizens in Tsarist Russia. When the Bolsheviks seized power they did not recognize the concept of nationalities and nationalism. Instead, their philosophy invoked the concept of class warfare, and Magda and her family, being

Revolution!

The political situation in Russia was becoming increasingly unstable. Russia's poor financial position was worsened by the disordered manufacturing and transport system, and by the devastated food industry: during the war the Russian countryside lost almost all of the healthy men including teenagers. In November 1916 it was clear that political upheaval was inevitable. The 1917 February Revolution resulted in overthrow of the Russian Tsar Nicholas II and the creation of the Provisional Government. It failed to hold on to power after the 1917 October uprising, when the Soviets led by Bolshevik party seized control and began the creation of the Red Guard. This was the beginning of Russian civil war that lasted for five years. The bloody fighting between the Soviet Red Guard (later to become the Red Army) and the anti-Soviet 'White' forces gradually moved from the central cities of Russia through the Urals and into Siberia. As a result of the final Bolshevik victory the Soviet Union was established.

Death of the Last Tsar

Tsar Nicholas II and his family were shot by the Bolsheviks in the basement of Ipatiev's house in the centre of Yekaterinburg on the night of 17 July 1918. One week later Yekaterinburg was taken back under the control of the White Army.

members of the gentrified classes, were deemed to be '*enemies of the proletariat*'. With the widespread implementation of the 'Red Terror' against all enemies of the new Bolshevik regime — both real and imagined — Magda's family were in terrible danger.

As the revolutionary battles rolled over the Russian countryside further to the east of Moscow, Yekaterinburg was caught in a terrible 'tug-of-war', repeatedly falling to one or another of the 'Red' Bolsheviks or the 'White' forces between 1918 and 1920. The situation in the city was one of utter anarchy.

This is how Magda Stromberg described these years in her diary on the 7th of November 1920:

> *'I have not touched this diary for the past three years. Many events both inside and outside of the country have made the lives of the people ever more complicated. The current circumstances of life differ so strongly from those previous to this. The Tsarist Government was replaced by the Duma; then Kerenskiy rose only to be replaced by the Bolsheviks.*
>
> *Kolchak[6] rules over the Urals during this time, and now all of Russia up to the Irkutsk is under the sway of the Reds.*
>
> *We are living on the edge of a volcano. There is no trade and at present we are desperately struggling to find enough food to save our children. There were times throughout these past years*

[6]The head of all the anti-Bolshevik White forces; formerly a Russian naval commander and Polar explorer, Kolchak's regime restored Tsarist laws, restored private ownership (both of land and factories), dissolved trade unions and persecuted Marxists. The regime was brutally repressive; in Ekaterinburg in excess of 25,000 people were executed.

when my sister Ella and I fed the family, both of us together, each working on two jobs, barely spending any time at home. The Estonian, Minna Kelman, took care of our children. Our men recently returned home, but this did not make things any easier. We live as one family on the Zlokazov dacha.'

Ella Kler, Magda's sister, died in August 1921 whilst caring for her husband who was ill with typhoid fever. At times, Magda was the only adult able to earn money for the upkeep of the four remaining children. The lives of the men in the family were hard too. As opposed to Ella, her husband, Modestus Kler, survived, but being a freethinking man in a totalitarian state the authorities constantly persecuted him.

Modestus O. Kler (26.12.1879–10.09.1966) was the son of a famous Ural geologist, Onisim Kler. Modestus also became a geologist as well as studying local history. Upon leaving secondary school in Yekaterinburg he spent nine years studying in Switzerland. Modestus studied natural sciences at the University of Neuchâtel and in 1904 he became a Doctor of Science at the University of Geneva.

The introduction of the New Economic Policy brought with it an influx of foreign investors to the Urals, seeking to start their own small businesses. Being a good socialite Modestus built up numerous contacts with the French businessmen that arrived in Yekaterinburg in the 1920s. As a result of his networking activities with these foreigners, Modestus

Stalin Comes to Power

In **1922** Joseph Stalin occupied the newly created post of General Secretary of the Communist Party. In this purely administrative post Stalin became highly influential as he effectively dictated the personnel policy in the party. As a result, in the last years of Lenin's life, members personally devoted to Joseph Stalin occupied all the key posts in the party. This left no practical alternative other than Stalin to be appointed as the new leader of the Communist Party. Stalin was named 'Man of the Year' by TIME Magazine in 1939 and 1942.

The New Economic Policy

Between 1921 and 1928 the New Economic Policy (NEP) replaced the earlier policy of Military Communism. Lenin introduced the NEP on 17.03.1921 initially as an attempt to prevent armed revolts by starving peasants. Cereal taxes were reduced, giving the peasantry a chance to recover after the dual ravages of World War I and the Russian Civil War. Later the currency reform of 1923–1924 introduced a new form of hard currency, 'Tchervonets', based on the gold index; certain elements of free trade were legalized, allowing for widespread cooperation between small businesses. At the same time, all medium to large industrial organizations remained in the possession of the government (they had been nationalized in the first few years following the October Revolution).

was twice tried on false charges of espionage and exiled from 1923–1925 and later from 1930–1931.

One episode in the life of another man in the family, Magda's second husband Alexander Vorobiev (27.08.1887–03.03.1949) also reveals the dangers faced by the unwary at this time. Alexander, a former warrant officer in the Tsarist army, was demobilized in February of 1918. He was a graduate of Moscow State University in economics and law, and subsequently became the head of the statistics department of the Yekaterinburg agricultural cooperatives. However, this was only to be a brief respite as on the 3rd of November 1918 he was mobilized back into the retreating White Army of Admiral Kolchak. During the retreat, Alexander fell ill with typhus and was left in a hospital in Omsk.

After he had recovered from his illness Alexander returned to work in his old post at the statistics department. This was at the time of the NEP, when the collective farms co-existed alongside private businesses, allowing both state and private farms to hire workers. Reasonably, Alexander and his colleagues compiled statistics concerning the productivity of both models of farming. They finally published a book that clearly showed that the private farming model was more productive than the *kolkhozes*.

The First Five-Year Plan

The New Economic Policy was abandoned in **1928** when the First Five-Year Plan for the development of the national economy was approved. It began an era of mass industrialization and collectivization. In order to maintain the extremely rapid pace of these transformations, a mechanism of repression was adopted as the universal solution for the extermination of both political and class enemies, and for the intimidation of ordinary people. Around 10 million people were effectively enslaved in NKVD camps throughout the USSR between 1921 and 1953.

However, when the First Five-Year Plan was adopted in 1928, Stalin and his supporters were pushing for the idea of total collectivization. So when total collectivization started on the 27th of December 1929, the results of Alexander's study and the book itself became unacceptable to the party leaders. As a result, the authors of the book were tried and sentenced to imprisonment in a labour camp or GULAG. Fortunately, these were still quite 'moderate' times, and so Alexander was freed from the camp after just one year, but was obliged to live and work in the small town of Revda in the north of the Ural region until his discharge in 1936. Alexander died in 1949, but not before he and Magda had two more daughters, Svetlana Vorobieva (1925–1926) who died while still an infant and Marina Vorobieva (24.06.1928–?).

Educating Armin

Soon after his arrival in Yekaterinburg, the matter of Armin's education arose. He was seven years old, which at the time was the usual age to enter primary school in Russia. However, no schools were open due to the continuous change of laws, the general lack of funds and the dire state of the economy. Yet as a Russian proverb says, 'poverty is skilful in ideas' (*голь на выдумку хитра*) which is similar in meaning to the English phrase 'necessity is the mother of all invention'. Magda eventually did

Education

Before the 1917 Revolution primary school education was organized both by the *'Zemstvo'* — the local government — and by the church in the form of parish and Sunday schools. From February of 1918 the new Bolshevik government tried to create a structured and centralized education system. They issued a decree transferring all schools into the possession of the Soviet government. The creation of new Soviet schools was not easy due to problems of ineffective management, poor conditions, and the ongoing civil war. The Russian famine of 1921–1922 closed four out of every five schools. Universal elementary schooling was only introduced at the beginning of the 1930s during a period of mass industrialization.

find a private school for Armin. This is how he recalls this time in his later memoirs:

> *'After the 1917 revolution the Tsarist schools were closed down and the new Soviet schools were only slowly beginning to be created. Mother made the decision to train me in a private primary school (for two years), organized by the Baev sisters in their home. They were former teachers at the old elementary school. The classes were held in a small house in the centre of Yekaterinburg. My classmates were mainly boys from the scientific and engineering intelligentsia families. Prior to the revolution the Baev sisters taught grammar, creative writing and mathematics in the tsarist gymnasium for first year pupils.*
>
> *After two years I continued my education along with my cousin-brother, Modestus,[7] and the son of Dr. Syano in the apartment of this doctor. Poluzadov, a former teacher of mathematics in the secondary school, trained us in arithmetic and grammar. At the age of 12 I fell ill for half a year with a severe form of scarlet*

[7]The son of Modestus O. Kler (Sr), the husband of Ella Kler, Magda's sister.

fever along with many other complications. Thus I lost a year of school along with Modestus. Yet, after only four years of primary education in those two private groups, I entered a new Soviet school straight into the sixth form instead of the fifth. Then I finished the last two years of school (the eighth and the ninth) in the newly opened experimental-demonstration school, named after Lenin. At that time the American method of education named the "Dalton-plan" was trialled in this school...'

'As soon as I finished school at the age of almost 17 (in 1927), the problem of my higher education arose. In those years this was really a problem, since only "socially close individuals" could enter an education institute, and the only option for me was to go through a school for workers (рабфак, "Worker's Faculty"), of which social class I was not a member: my parents were employees and their nationality was German. A solution was found in the fact that my mother worked as a tutor in the Ural Industrial Institute (UII). This determined my specialty to a certain degree — as a chemist. With the help of the Institute, permission was obtained from the Education Ministry to accept my entry to study there.'

It's interesting to read Armin's first-hand account of what the higher education system looked like after the first ten years of Soviet Russia under the new regime:

'In 1927 I enrolled in the chemistry course in the chemical-metallurgical department of the Urals Industrial Institute (UII). At that time only 25 people took a course in chemistry at the chemical department (one group). The people in the group were a varied lot. Apart from myself — a 16 year-old boy — our group included a former worker's faculty student, Krukov, who was approximately 30 years old and a participant of the civil war, and a female interpreter from a scientific research institute who was approximately 40 years old. During the third year of my study in UII (in 1929), due to the commencement of the First Five-Year Plan (in 1930), the slogan "success is determined by the amount of qualified

personnel!" was proclaimed and an order for "accelerated graduation" was issued to all students in the education institutes. As a result of this I only studied for three and a half years instead of five. In May of 1930, at the age of almost 20, I obtained my graduation 'certificate' from UII in chemical engineering (specializing in electrochemistry). Of course, my diploma thesis was not written or submitted, so I had only a poor grasp of any technological disciplines. Fortunately, my lack of technical knowledge was not necessary in my future life as I moved straight into a scientific research institute and never worked as an industrial chemist in a plant.'

After graduating from the institute, Armin worked as a junior researcher in the analytical laboratory of Magnesium, headed by Professor Scherbakov at UNICHEM, a scientific institute supporting the chemical industry in the city of Sverdlovsk.[8] He then moved to a new institute in the city:

'For two years I participated in producing the first kilograms of metallic magnesium via the electrolysis of fused salts from a raw material named carnallite found in the Urals (from a deposit near Solikamsk).

In 1932 I transferred to the newly created Urals Physical Chemistry Institute (URALFIZKHIM, Sverdlovsk city), as a junior researcher working in the laboratory of the electrochemistry of fused salts, which was led by senior researcher S. V. Karpachev.'

The USSR Academy of Sciences generally enjoyed a period of rapid expansion and progress between the years of 1921 and 1934. Four years after the Revolution Russian scientific activity had restarted by the end of 1921, with the first scientific congresses convened in Petrograd.[9] This

---

[8]Modern day Yekaterinburg — the name was changed in 1924 in honour of Yakov M. Sverdlov, a prominent Bolshevik, who died prematurely during an influenza epidemic in St. Petersburg in 1919. He was credited with organizing, remotely from St. Petersburg, the fate of the Romanov family in the then named city of Yekaterinburg.

[9]The name was given to St. Petersburg between 1914 and 1924 in an effort to Russify the name and avoid the obvious Germanic connection.

'rebirth' coincided with the adoption of the NEP and with the decree (issued on the 6th of December 1921) by the Council of the People's Commissars (the Soviet government) concerning the improvement of working conditions for Russian scientists.

Alongside the programme of intensive industrialization that followed the adoption of the First Five-Year Plan in 1928 the Academy of Sciences became of utmost importance for the Soviet government. Major changes in the structure and management of the Academy occurred. These included the appointment of a new board and also the introduction of a new commission dealing with the planning of research. This effectively made the Academy a servant of Russian industry. At the same time new branches of the Academy were opening in different parts of the USSR, such as the Ural branch (opened in 1932) mentioned above by Armin. Nevertheless, things were about to change from 1934 onwards.

Walking on the Edge of a Precipice

'Walking on the edge of a precipice' was a phrase often used by Armin Stromberg to describe the period of his life from 1930 onwards and it is a fitting description of what was happening throughout the country during the second half of the 1930s. At this time Armin recalls that at first he was young and careless and that he did not fully appreciate the events unfolding around him, nor the dangers that they could hold for him and his family. The time of the 'Great Terror' was approaching, when a careless slip-of-the-tongue could easily attract the attention of the NKVD — often with fatal consequences:

> *'I am surprised how I managed to survive; according to the realities of the Soviet Union I should have been destroyed.'*

Armin attributes his survival to the fact that he tended to work alone, even after he had defended his PhD thesis, without any subordinates who might inform on or denounce him to the NKVD.

> *'I avoided any arguments because of the fear of denunciation. I was constantly alert, and society's suspicious attitude towards me also helped to keep me somewhat isolated.'*

The likelihood of being denounced was high even amongst ordinary citizens as the NKVD was constantly purging the 'undesirable' members of Soviet society on a planned and regular basis. Sadly this process was often exploited for personal advantage by younger, unscrupulous or ambitious colleagues in the workplace or co-inhabitants of a flat so as, for example, to gain a promotion (filling the 'dead man's shoes'), or to occupy the entire flat for themselves. All that was required in order to have one's neighbour or work colleague sentenced was to write a denunciation accusing them of any imaginable political crime. Such 'crimes' even extended to simply telling a joke or an impolite anecdote about the Party or any of

Ezhovshchina — The Great Terror

On the 1st of December 1934 Sergei M. Kirov (the head of the Communist Party in Leningrad and a powerful opponent to some of Stalin's more extreme policies) was assassinated in Leningrad, most probably on the instruction of Stalin and in collusion with the NKVD. His murder heralded the introduction of extraordinary arrangements and the implementation of a reign of terror against the Old Bolsheviks and ordinary people (not just the peasantry) alike. On the evening of the same day as Kirov's murder an edict was issued introducing 'accelerated judicial procedures' that brought about the immediate execution of any death sentences conferred on prisoners. Over the next several days many of the so-called 'White Guards' and political prisoners were summarily shot in Leningrad, Moscow and Kiev without the due process of holding a trial in a court of law. This marked the official start of the widespread campaign of state terror. On the 25th of September **1936** Joseph Stalin and Andrei Zhdanov sent a telegram from Sochi, ordering the replacement of Genrikh Yagoda with Nikolai Ezhov as the head of the NKVD. The following two-year period of state terror that peaked in 1937–1938 with a lower estimate of over 680,000 victims shot — an average of more than 1000 per day — came to be known as '*ezhovshchina*' after his surname, or The Great Terror.

its leaders. Given Armin's German nationality, the people around him were somewhat reluctant to associate with him (or even to mention him by name) as they too feared denunciation as a traitor, spy or collaborator. Bizarrely this social exclusion also had the unexpected benefit of reducing the chances that his peers could falsely accuse him of a crime!

Scientific Isolation

This febrile atmosphere of mistrust and paranoia was certainly not conducive for producing world-class science, as any individual that became too prominent or successful was almost certain to end up being shot. Yet, in spite of this very real risk some of the old-school scientists did continue to have their research printed in foreign, international journals published abroad. They still imagined that they could work as they had prior to the Revolution, where science knew no borders. Any scientific freedom that was afforded to them was abruptly ended after the Luzin affair of 1936. This led to a major campaign by the authorities to clamp down on scientists who chose to publish their work outside of the USSR. From that time, and for many subsequent decades, Soviet scientists ceased to publish their work abroad, and Soviet science became totally isolated from the rest of the world. This was to have a profound influence on Armin's scientific work — only very late in his life did he begin to dare to publish his findings in foreign journals — and is primarily the reason why, until now, his pioneering and hugely influential work has past barely noticed by chemists, and in particular electrochemists, outside of the former USSR; but we are jumping ahead of ourselves. Let us return to the events of Armin's life in 1937.

3

Letters from the GULAG (March 1942–September 1943)

In 1937 Armin G. Stromberg became engaged to Lydia Mikhaylovna Poponina (1912–1962). She was the daughter of illiterate parents who served the landed aristocracy of Shevelin, a region in the Southern Urals. Her father, keen to take advantage of the new opportunities for the poor in post-Revolutionary Russia, had encouraged all five of his children to gain an education. Armin and Lidusya[1] were married upon the completion of her physics degree from Leningrad University that same year. In Armin's own words:

> *'To me it was necessary to have a close person, with whom I could share sorrows and joys and on whom I could rely in case of emergency... After two months practice, several days before her departure to Leningrad, I proposed. She agreed. This was an act of folly on her part.'*

The wedding was modest, reflecting the relative poverty of academics in the Soviet Union at this time:

> *'Without an immense banquet, without Mendelssohn's Wedding March nor any riding in motor vehicles with coloured balloons.'*

[1]An affectionate form of Lydia's name commonly used at home and by members of Lydia's family.

realized that the deported Germans would make for a very useful and highly productive workforce for the war effort. He therefore took a personal interest in effectively enslaving the deported Russian-Germans into what became known as the 'Trudarmy'.[2]

The order of 'transportation' could seem tantamount to a death sentence with an arduous and dangerous journey followed by the harsh living conditions in Siberia. In one documented instance, 400 children died in a single freight carriage used for the journey to Siberia. Added to this was the possibility of Soviet agencies being used to execute prisoners. Earlier that same year on the 28th of July 1941, 80 prison officials were transported from Engels to Kursk to execute members of the transported Volga-German population.

Around this time a message was received which directed Stromberg 'to come next day to the police with the passports of the entire family' for deportation to the Northern Urals. In this instance the Strombergs were considered political undesirables on account of their German ancestry. However, by good fortune — and his mother's connections — Armin Stromberg and his family avoided deportation in 1941. Upon receiving the message to report to the police, Armin's mother, Magda, went directly to the rector of the Ural Polytechnical Institute (UPI) to plead their case. The rector, a promoted worker who had risen through the ranks of the Communist Party, was well known and respected by the local party leaders. He was thus able 'by the right telephone call' to prevent the Strombergs being sent en masse to one of the closed northern cities.

> *'He was a very curious rector, who had neither degrees nor titles. However, he was the former member of the regional committee of the Communist Party of the Soviet Union. In time, he had been directed to work at UPI. He was a well-connected, likeable person. He instructed Magda not to go anywhere — and this was made possible by the right telephone call.'*

---

[2]Trudarmy (*rus. трудармия*) — literally 'working army', was the public name given to the secret system of concentration and forced labour camps. Most of the Trudarmy 'soldiers' (prisoners) were members of minor nationalities, such as the Germans or Tartars.

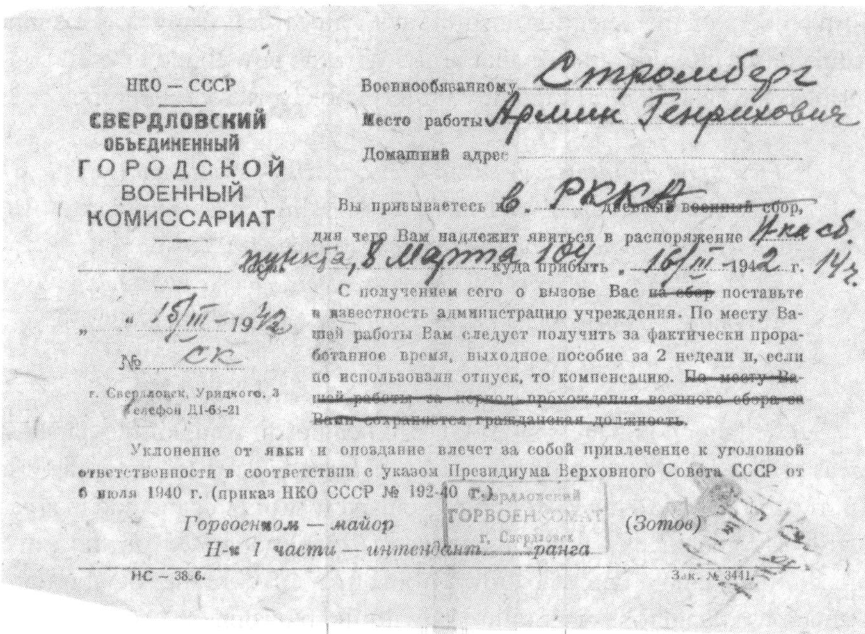
НКО — СССР

СВЕРДЛОВСКИЙ
ОБЪЕДИНЕННЫЙ
ГОРОДСКОЙ
ВОЕННЫЙ
КОМИССАРИАТ

„15/III-1942"

№ СК

г. Свердловск, Урицкого, 3
Телефон Д1-65-21

Военнообязанному Стромберг
Место работы Армин Генрихович
Домашний адрес

Вы призываетесь в. РККА ~~дневный военный сбор~~,
для чего Вам надлежит явиться в распоряжение Нка сб. пункта, 8 Марта 104 куда прибыть „16/III-1942 г. 14ч.

С получением сего о вызове Вас ~~на сбор~~ поставьте в известность администрацию учреждения. По месту Вашей работы Вам следует получить за фактически проработанное время, выходное пособие за 2 недели и, если не использовали отпуск, то компенсацию. ~~По месту Вашей работы за период прохождения военного сбора за Вами сохраняется гражданская должность.~~

Уклонение от явки и опоздание влечет за собой привлечение к уголовной ответственности в соответствии с указом Президиума Верховного Совета СССР от 6 июля 1940 г. (приказ НКО СССР № 192-40 г.)

Горвоенком — майор Свердловский ГОРВОЕНКОМАТ г. Свердловск (*Зотов*)
Н-к 1 части — интендант ...... ранга

НС — 38.6. Зак. № 3441.

The fateful order of mobilization received by Armin in 1942 which reads:

To the citizen liable for call-up: Stromberg Armin Genrihovich

You are called in RKKA (Workers' and Peasan' Red Army), for which you are required to appear at the disposal of the chief of assembly station in 104 of 8-March street on 16.03.1942 at 14:00.

Upon receipt of this call, you must inform the administration of your organization that you are called for ~~training~~ assembly. You should obtain your salary for your time spent working in the place of your employment, mustering out pay in two weeks and, if you have not used your full holiday pay, then obtain a compensation. ~~Civil post is kept for you during the period of the passage of military assembly.~~

Conscription avoidance or delay entails the penal responsibility in accordance with the Decree of the Supreme Soviet of the USSR dated 6 July, 1940. (order of the NKO USSR No. 192-40)

Subscribed by: the city military commissar Major {stamp instead of signature} (Zotov)

This respite was temporary. A few months later on the 20th of March 1942, as German forces began the battle of Stalingrad and the siege of Moscow continued unabated, Armin received news that he was to be mobilized into the RKKA (the Workers' and Peasants' Red Army). Like all members of the scientific institute with a degree, Armin was exempt from conscription into the regular army. What he now found himself conscripted into was a 'working army' of (effectively) slave labourers.

> *'First of all this army was called a 'secret army' — one could not find any information about it in the press. It was decided to isolate all Germans under the pretext that they were the spies and agents of Hitler.... No documents relating to this decision were published, everything was performed on the unofficial secret instructions of the NKVD.'*

These labourers were sent to GULAG camps in some of the harshest areas of Russia to work, producing the necessary raw materials for the war effort. Iron and coal mines were run exclusively using this method in areas of Siberia and many munitions and arms plants operated on the same model. Slave labour could be drawn from any number of the social groups deemed as 'dangerous' by Stalin. As with the punishment battalions that would later be formed by the Red Army, it was not a priority to keep these workers alive and the survival rates at some of the more exposed camps, particularly those in Northern and Eastern Siberia, were truly appalling. In Armin's instance it was the German inhabitants of Western Siberia who were rounded up to form the 'German Special Force 18–74' destined for the GULAG camp 129 km away at the closed city of Nizhni Tagil to operate the brick plant there. Of the 6000 'soldiers' originally conscripted, only 3000 would remain after three years of working at the camp.

> *'Nizhni Tagil was not simply one camp surrounded by barbed wire. It was a system of several tens of work camps attached to different construction plants. Prisoners were not allowed to go near the important machine* [or vehicle — GGW] *assembly plants ... they felt that the Germans were not to be trusted. We were attached to a brick factory. The brick plant together with its associated camp was called 'Special Camp 18–74' and together the entire complex of individual camps was called 'Tagillag'. In the camps were held not only Germans, but also ordinary criminals. For example, there was one section of our camp whose barracks*

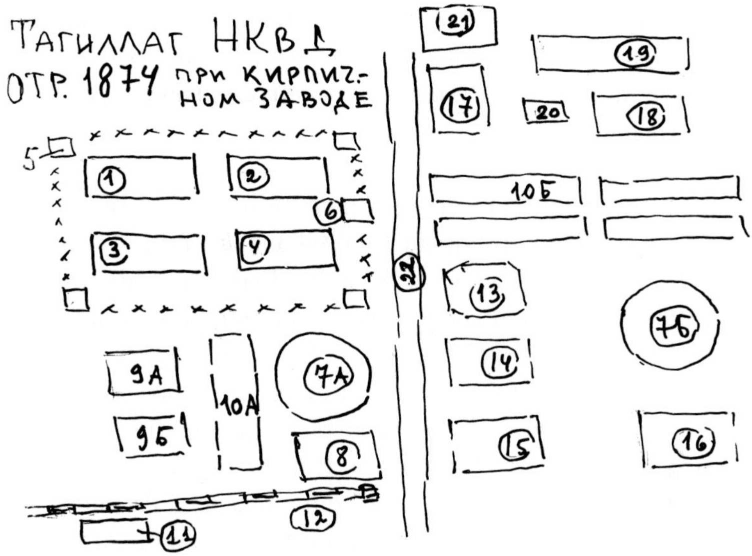

A sketch drawn by Armin showing the layout of the brick plant and barracks of Special Force 18–74 between 1942–1943: (1–4) Four barracks where the ‘soldiers’ of Special Force 18–74 lived. The German engineers (ITR) lived in barrack house no. 4; (5) Enclosure of the barracks with barbed wire with watchtowers situated in the four corners; (6) The clock house; (7A) The working Byurera kiln; (7B) The inactive Byurera kiln; (8) The inactive Hoffman kiln; (9A–B) Two buildings of the brick-moulding workshop; (10A) Winter apparatus to dry raw brick materials heated from the flue gases of the Byurera furnace; (10B) Summer apparatus to dry raw brick materials without any special pre-heating consisting of covered blast-proof regimental sheds; (11) The loading shop; (12) A sliding-rail (blind alleyway) for the dispatch of the finished brick products (onto railroad wagons or platforms) or for inspecting the fuel (fine coke and bituminous coal) for the Byurera furnace; (13) The brick plant’s club for the workers of force 18–74; (14) The bakery; (15) An office block; (16) The dining room; (17) Workshops (mechanical, carpentery, pottery, cobblers and tailors); (18) The garage/motor pool; (19) The hospital/infirmary; (20) OTK; (21) The clay quarry; (22) Armin did not provide mention of this, although it seems likely to have been the road linking the various camps to the city.

> *were located near to the clay quarry (called ‘Zay’ — or ‘Hare’ — mountain), where the Germans obtained clay from which to make bricks. Such branches* [of the camp, in this case Special Camp 18–74 — GGW] *were called ‘missions’. Within such missions*

> *they frequently carried out specific tasks, such as storing hay, cultivating agricultural land in order to feed the camp, etc. In this way, the 'Tagillag' conglomeration comprised of approximately 50 such camps.'*

We have the unique privilege of gaining a detailed insight into the daily life of prisoners in the Nizhni Tagil GULAG complex via the 74 surviving letters that Armin wrote to his wife, Lydia, and his four-year-old daughter, Elza, in the months between his internment in March 1942 and his eventual release in September 1943. Unfortunately, the replies from Lydia no longer exist, but the content of these letters, and Elza's recollections of this early period of her life, also provide us with some description of the deprivations suffered by ordinary Russian civilians, separated from their loved ones by war and unfortunate circumstances. Armin's story, told in his own words in these letters, is a remarkable tale of suffering, isolation, fear, insecurity, friendship, humour, hope and the capacity of a remarkable man to turn his brilliant scientific mind to ensuring matters of his own personal survival, and that of his friends and comrades around him. The condition and scarcity of food becomes a major theme in his letters, with a precise, analytical description of the amount of rations and their calorific and monetary value occurring in almost every letter; but what is even more apparent is that this primary concern with food extends not only to him and his fellow inmates, but also to his wife and child. There is an underlying human story in Armin's letters, of the love and the pain of separation felt by a husband for his wife and young daughter, and also the fear of abandonment as evidenced by Armin's frequent pleas to Lydia to write more often. As Armin notes in later letters it was not uncommon for mobilized husbands to discover that their wives were unable or no longer willing to cope with the prolonged, forced separation and requests for divorce were not unheard of. Throughout these letters Armin plays down the difficulties he faced whilst at Nizhni Tagil, and constantly attempts to reassure his loved ones that all will be well, whilst he himself was fighting for his very survival amidst difficult and dangerous circumstances.

We, the authors, have chosen not to abridge the following letters with superfluous commentary. Nor have we succumbed to the prevailing trend in modern historical texts of pointing out to the reader what Armin himself makes patently clear, except to add explanatory footnotes or additional facts from other sources of correspondence such as the few surviving fragments of Lydia's replies to her husband, or Elza's own recollection of events and family relations.

So, we now hand over to Armin himself, who begins the story of this period of his life with his first letters upon arriving at the train station in Nizhni Tagil, wondering what the future has in store for him.

Letter 1

Friday 20th March 1942
(Nizhni Tagil, the railway station), 12.00 p.m.

Dear Lidusya!

Here we are in Nizhni Tagil. Our journey was quite good. We began boarding 15–20 minutes after you left. They gave us a third of a third-class railroad car, no. 1 in train 21. The coach was warm. I sat at a second berth; Berngardt[3] was above me in the luggage rack. I also kept the third berth for Ryukkert.[4] We had supper using boiling water, which was given to us by a conductor. We are named as 'the command 36a'. We arrived in Nizhni Tagil at eight in the morning, and up until this point we are sitting inside the station, which is warm. Probably, we will soon be moved further on. They fed us with a working dinner — two clay earthen saucers of soup with dumplings and 200 g of bread, so that now I'm in perfect bliss and even ready to walk 20 km if needed. What lies in front of us?

I will write to you as soon as I look around our destination. What does my daughter say about me? Kisses.

Your working soldier.

Regards to everyone. Pass on my thanks to Mum for the rissoles.

---

[3]An artist and friend, the son of Wanda Antonovna, a music teacher — EAZ.

[4]Not to be confused with Rickert later [EAZ]. Armin tells us about him later in Letter 3.

Letter 2

Saturday 21st March 1942, 10.00 a.m.

Dear Lidusya!

Here I am and almost at the place of my designation. Almost — because today they will sort us by our speciality and we will settle accordingly. All this must be resolved during the present day. Thus far I cannot report any special news about myself.

Yesterday the whole day was spent travelling. From the station they moved us temporarily into barracks, where we drank boiling water and then I slept sweetly for one and a half to two hours on a cot with a mattress. At three o'clock in the afternoon they met us, and guided us to the tram. We left the city on that tram and camped near the tram stop whilst the commanders figured out where to go next. After two hours they realized that we had not come by the right route. Fortunately they found a truck for our belongings. We went in the opposite direction on the tram No. 2, arrived at the city centre and then walked 5–6 km on foot to the brick plant. Here the chief assistant of the special column made a roll call, and settled us into some temporary accommodation with bunks on three levels. [See the picture of letter five given below — GGW.] I've occupied the middle bed near the window, but it was not too cold to sleep, since the people on duty hauled pieces of wood from around the barracks and lit a stove. In the evening we drank boiling water, which people brought from a cauldron 200 m away from the barracks. But they must give us a water tank and then all this should be arranged. This morning they gave us 600 g of bread for the day (thus far we have not worked). I divided a loaf into three parts: to Berngardt, to Ryukkert and to myself. Also they brought a pot of skilly.[5]

The people in our 'command 36' are very good: they can take initiative and are disciplined. We immediately found a broom, and began sweeping; we also found a saw and used it to cut some firewood. We

---

[5]In Russian: *balanda* (баланда). A thin soup or gruel consisting of water and peas — GGW.

found boards for additional bunks. I do not worry about the possibility of theft of my belongings if we stay together.

There are approximately 400 of the same 'specialists' in our column. The chief informed us that the name 'special column' means that in this column all different specialists from the unskilled worker to the engineer are gathered. Ryukkert already met one familiar specialist — a turner, who moved to the Garinskiy region in autumn with his family, and was mobilized in RKKA (Workers' and Peasants' Red Army) from there, just the same as I was.

The speciality of everyone was recorded this morning. There are engineers among us: a heat engineer, Scheurer (he will sleep above me), an engineer-designer, two mechanical engineers — Motte and Krechman. Also there are turners, metal craftsmen, tractor operators, electro-boilers, etc. There is one miner — Geer. The question of our future place of work and our speciality is the main concern for us now. Today this decision should be made. A barber has come to cut our hair (I already shaved this morning). The commissioner will soon arrive to talk with us.

What does the future hold? Time will tell. It is already clear that the turner will find a skilled job for himself more easily than a chemical engineer.

Wholly yours, Lidusya. Kisses from me to Elza. Does she attend kindergarten? Regards to all our relatives. Inform Modestus.[6]

Your labourer soldier.

Letter 3

Sunday 22nd March 1942, 5.00 p.m.

Dear Lidusya!

Presently the third day of my stay in Nizhni Tagil is drawing to a close. I sent the first postcard to you from the station. Yesterday I sent a letter with a description of my first impressions of my new place of residence.

[6]Modestus Kler (1910–1963) — Armin's cousin and the son of Modestus O. Kler, (Armin's uncle interred earlier for three years for alleged espionage) — EAZ & GGW.

My future work prospects were narrowed down today, but this is little consolation to me thus far.

Berngardt and I worked our first shift overnight from the 21st to the 22nd of March with a brigade of 12 men: unloading trolleys of baked bricks from the kilns and sorting them into piles. These furnaces have the form of a tunnel. The temperature in them is like that of a bath. We warmed ourselves in them in the breaks between unloading. The work is unpleasant, since the bricks are very dusty and everyone was terribly soiled. We changed shifts after eight hours with a break of one hour for breakfast. We will be living, probably, in the same room with 30 other people. I, Berngardt and Ryukkert are part of the seventeenth section of the twentieth column consisting of 24 people under the command of Krechman.

Our daily routine: 4.30 a.m. reveille, toilet until 5.00 a.m., breakfast until 6.30 a.m. — (skilly and bread), work begins at 7.00 a.m. (eight-hour shift), dinner at 6.00 p.m. — skilly, herring and bread — evening roll call at 8.30 p.m. and lights-out at 9 p.m. To date they have given us 600 g of bread each, and generally this depends on the productivity compared to a plan or 'standard': for achieving 100% they give us 700 g, 110% — 800 g, 120% — 900 g, 90% — 650 g, 80% — 600 g, 70% — 550 g, and so on until 40% — 400 g. If you do not work, for example if there is no footwear etc., than you get 400 g of bread. Gradually people are finding jobs related to their speciality. Rympel became a metal craftsman in the auto-garage; Hoff, Pekher, Soor became joiners; Mayzinger (a plant manager) became the assistant to the chief of one of the columns; Rivelman took the post of inspector of educational work; Berngardt and Baumann (heat engineers) do the mass cultural work. I have not been assigned anywhere so far. I sent a letter to the chief of the central laboratory of plant 183 about my work on the polarograph.[7] I wanted to speak today to the chief engineer of the brick plant about the possibility of employing me at the plant based more or less on my speciality, but there is little hope. In any case the assistant to the construction chief said that I'm already the eighth chemist in the

[7]The 183rd enterprise was formerly the Kharkov locomotive plant no. 183, which was dismantled ahead of the Nazi advance and shipped to Tagil to become the main manufacturer of T-34 tanks — ASK.

construction crew[8] and not one of these got a job according to his or her speciality. By the way, I met with a certain Varga in the office waiting room of the chief engineer. In December (1941) he was moved to the town of Gori with his family (his wife and eight-year-old child), and from there he was mobilized a few days ago. He is a bookkeeper and the chief engineer promised to find a job for him.

That's how I live, Lidusya. Thus far it's possible to live and to walk. By the way, the other day they permitted soldiers to receive parcels. The address for these parcels is: Nizhni Tagil, Sverdlovsk region, *Tagilstroy*[9] of NKVD, construction crew 18–74, Stromberg A.G. Lidusya, don't send anything for now. Perhaps you can send something small (generally it's permitted to receive parcels up to 8kg) in order to verify if the package will arrive. Tomorrow I will try to talk to the chief engineer and then I will send a letter to Grigoriy Ivanovich. With impatience I await a letter from you. Please tell me which letters you have received from me and when. Write about our daughter, about all our relatives.

Your building-soldier.

Address for letters: N. Tagil. Post office 2, Postbox 67.

Letter 4

Sunday 23rd March 1942, 5.00 p.m.

Dear Lidusya!

I sent a postcard to you from the station … a letter with my first impressions … a letter about my first day of work stacking bricks from trolleys into piles…

Today the work was harder. From 7.30 a.m. to 12.00 p.m. we cleared snow from a layer of coal along the line of railroad; then, from 12.00 p.m. to 4.30 p.m. we unloaded a railroad car (60 tons) of coke with shovels. We unloaded about two-thirds of this and surely they will not consider this as

[8]Also well known as transliterated Stroyotryad from rus. Стройотряд.

[9]Tagilstroy (rus. Тагилстрой) A department of the NKVD overseeing building operations in the Tagil region.

meeting the standard daily plan. Thus far I still have some bread left from home, so it's not too bad, as it is still possible to get 400 g of bread and survive. Yesterday, in the evening, 20 more men of German descent — the collective farmers — moved into our room. There was a large crowd resulting from this movement of people. Our section (24 people) was moved into the part of the room where I slept, so I did not have to bother moving. Part of our section (12 people) was moved to other rooms. The collective farmers took up the other half of our room. Now 44 people live in a room of 50 m^2 (5×10). Yet this is not so terrible. We all are very relieved that they did not move us into the very cold barracks, where several hundred of those who were mobilized now live.

I went today to the chief engineer of the brick plant Filatov. As I expected, the conversation did not go well for me. He said that he had recruited enough engineer-Germans, more than enough in fact (six people), and that chemists are no longer necessary, taking into account that there is no chemistry involved in brick production (there is no laboratory at the plant). Meyzinger (he is now the chief of the local political educational office) said to Berngardt that he will take him on in his staff and that there will be much artistic work for him. I also want to talk with the chief dispatcher of *Tagilstroy* about the possibility of using me in other plants, but I pin no special hopes on this conversation. The assistant chief of the building group told us, for example, that the other day all the Germans machinists were removed from their jobs and replaced with criminals.

How are your domestic affairs, Lidusya? Call Genet. Ask her to ask Al. Iv. about the money (100 + 300 roubles) from the machine shop for me. I will continue this letter tomorrow.

Tuesday 24th March 1942, 4.00 p.m.

Yesterday in the evening I had a chat with Veybert. He was born in Sverdlovsk. His parents were born and lived in Russia. His family has lived in Sverdlovsk since 1900. In January they[10] wanted to move the family, but his father needed a surgical operation, so they stayed due to his father's illness. In the autumn, young Veybert had entered the mechanical

[10]The NKVD — ASK.

department of UPI, but then he took a job as a turner in a military plant. When they mobilized him, the plant pleaded with the military registration and enlistment office to let him stay, but they were unsuccessful.

Today we again worked from 7.00 a.m. to 3.00 p.m.. First, myself and three others out of the ten of us cleared snow from an area used for piling the bricks. This work is not difficult and clean. In the last three hours, we four were joined by the remaining six men, and we picked out the broken brick — 'halves'– with crowbars from the heaps of rubbish and snow, sorted the bricks into separate heaps, and then loaded them onto a lorry. In the few moments of leisure we had, we warmed ourselves near the bonfire inside the brick kiln. With the timid light of a bonfire, which illuminated the gloomy arches of the cave (these furnaces do not work in winter) a conversation started amongst us. Schumacher's great-great-grandfathers were shoemakers of Tsar Peter I (hence his surname — *Schuhmacher*). Alboni's father was Italian. When he was a boy, he went to Germany to study with a German family there. When he was older, he left his surname Alboni and accepted German nationality. So the Italian Alboni became a German.

Berngardt began his artistic activity today. He wrote inscriptions for the office doors: accounts, the chief of construction crew, etc.

Now there is a hurry-scurry in the room again. Make bunks. The population of the room must grow to 60 people. I feel myself healthy, cheerful, as if on a health resort, only the menu is a little unlike a health resort.

I worry about how you are coping without me. Have you obtained coal? Tomorrow (the 25th of March) Elza must go to the kindergarten. In my thoughts I'm with you and this comforts me.

My address: Nizhni Tagil, Post office 2, Postbox 67, Stromberg A.G.

Letter 5

Thursday 26th March 1942, 4.00 p.m.

Dear Lidusya!

My life is gradually progressing along a set path and, after a certain time I will become so accustomed to this life that I would not want for anything better. I will describe to you, for example, yesterday's events. At exactly

6.00 a.m. we get up from our beds and we begin to dress. We wash with snow in the street (there is a washstand, but it has not been installed yet); and we eat a pea skilly (Bernhard and I share one cup). Around 7.00 a.m. we leave for work (the brick plant is located about five minutes on foot from the barracks). The dispatcher then assigns work to us. Yesterday our brigade was assigned to the loading of bricks onto flatcars, and another brigade was requested to dig a grave. The flatcars that we were to load had been placed outside of the loading area. Thus everyone had to carry around five pieces of brick up to a distance of 20–25 m from the trolley to the loading platform. Our brigade of ten people loaded one and a half platforms, about 10,000 pieces of brick. Thus each of us made 1000/5 = 200 journeys of 25 m with the bricks and 25 m back. Furthermore, two men of our brigade carried out some other work. On the whole we performed the standard to 115% and earned 900 g of bread. At 3.00 p.m. we finished the job, came home and had a snack (I saved some herring from the previous day and after arriving home ate it with the bread).

Friday 27th March

I continue this letter, since I was appointed as the person on duty for the section after we received our bread ration last night. After this meal I sit down to write letters, then I darn trousers or make mittens. Someone brought a skilly along with one herring. Someone else volunteered to act as our postman and we all gave our letters to him as he left. Another man went to fetch water from a spring (there is a titanium water tank not so far away, but the water in it smells of carbolic acid). I boil the water on a plate and drink tea. Yesterday, finally, it was possible to read a little scrap of a newspaper. As the light fades in the evening we go to bed.

A *Stakhanov*[11] brigade of workers from Turinsk works side by side with us. The brigade leader, Izrukov, is the docent of the mining institute (in the department of the development of ore deposits, a former student of Shevyakova). I chatted with him. He knows Mum and Mitya Ivliev. In January he was re-exiled into Turinsk, where he worked

[11]Named after a prominent worker A. Stakhanov, whose name became a common noun for workers whose work exceeded expectations/orders.

before his mobilization as the head of the supply division on a state farm. His wife (who is Russian) continues to live with their children in Sverdlovsk.

Another guy, Eshman, is also in our section. He was a turner and was evacuated together with his plant from Leningrad. He had an exemption until May 1. When the call-up papers finally arrived, the factory management and the special office instructed him not to go to the military registration and enlistment office. But all their efforts proved to be useless.

People are being arranged gradually into jobs. Two men got the job of dispatchers in the construction crew. One person got the post of dispatcher for the entire plant, another person a task master, another one is a norm-setter. Six men became squad leaders. Only 12 of 36 unskilled workers remain. I attempted to arrange a job, but thus far with no success. I wanted to speak to the chief dispatcher of *Tagilstroy*, but his office is in the city, and they did not permit me to go there. I am no longer hoping to get a job as a chemist, and I want to attempt to be arranged at least as a technical worker in the machinery plant. I called in at the planning section and the bookkeeping office of the brick plant, but they don't need any more workers. They need an engineer estimator, but some construction knowledge is required for this post. Well, it is not a big deal; I will be an unskilled worker. My boots and mittens are intact. I'm dressed warmly (thanks to Tosha for the jacket). I'm healthy, and do not lose heart. My muscles ache, being unaccustomed to hard work, but soon I'll get used to it and everything should turn out fine.

Just in case, I will give you my address for any messages and parcels: Nizhni Tagil, *Tagilstroy* NKVD, construction crew 18–74, column two, Stromberg A.G. Try to send a small parcel as a test. Send a pair of tubes of calcex (I already ate my entire reserve), some onion, envelopes and postcards, anything that is edible. Do not send much. It is simply interesting to me whether the parcel will reach me. At the moment, no one in our department has received a single letter. I await one from you before long. Is everybody healthy?

Kisses — Armin.

Write the addresses of Modestus and Elga.

Letter 5A

Friday 27th March 1942

Dear daughter!

How are you? Do you go to kindergarten? Your dad is in the army now, but he does not shoot the fascists with a rifle or a machine gun, but makes

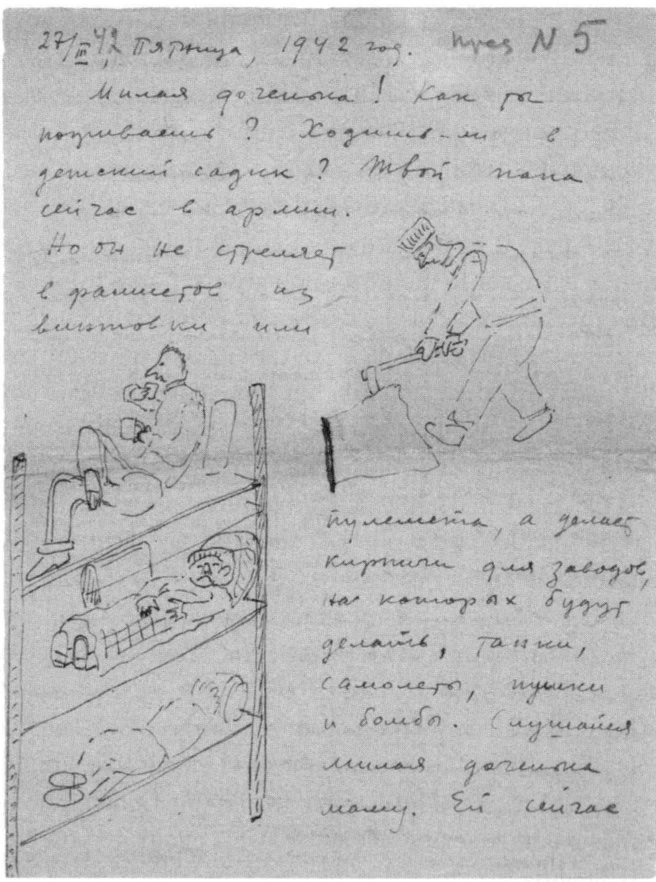

27/III 42 г. Пятница, 1942 год. [illegible] N 5
Милая доченька! Как ты поживаешь? Ходишь-ли в детский садик? Твой папа сейчас в армии.
Но он не стреляет в фашистов из винтовки или пулемета, а делает кирпичи для заводов, на которых будут делать танки, самолеты, пушки и бомбы. Слушайся милая доченька маму. Ей сейчас

A scan of Letter 5A written on a separate quarter of a page with two figures — one depicting three-storied bunks with three 'soldiers' — one eats while two sleep. The second shows someone picking out the broken brick halves by hand off a frozen heap.

bricks for the plants, at which the tanks will be made, as well as aircraft, guns and bombs. Obey mum, dear child. At the moment she has a hard time without daddy. Help mum. Dress by yourself, eat without any fuss. Go to the kindergarten. Do not fight with Viktor. If he offends you, then tell him that you will write a letter about this to your dad in the army. Write a letter to dad. Make pictures of a little fellow, tanks, aircraft with your coloured pencils. And let Viktor draw a picture for me. Tell Michael and Leo Bogdanovskiy that they could write a letter with some figures to me also.

Goodbye daughter. Father greatly loves you and wants to see you. But they cannot dismiss dad from the army yet, and therefore he is mobilized. When the war ends and they beat the fascists, dad will arrive home and will strongly kiss his daughter.

Your dad.

Letter 6

Saturday 28th March 1942, 4.00 p.m.

Lidusya!

I have lived in Nizhni Tagil for eight days, and it's the eighth time that I sit down to write a letter to you. My mates mock me because of my frequent letter writing, but I simply need it. In the process of writing these letters I seemingly converse with you, I return to our united life.

There was a snowstorm today, and our brigade cleared the snow from the area in front of the furnaces, railway lines, etc. For three days we have exceeded the standard and received 900 g of bread and a third cauldron (i.e. a thicker pea skilly and some fish). Furthermore, today they introduced a third daily meal: skilly and pea porridge. Now our brigade has returned home, we ate up all this food and are now resting in a good mood.

Yesterday, in the evening, four more people were added to our section. The Shildas — the father, 47 years old, and his son — are among them. It turns out I met the young Shilda a year ago on the skating rink (he ran the figure skating). They are political emigrants from Czechoslovakia (1939).

I have become much closer friends with Schumacher (50 years old), Veybert (19 years old) and Alboni than with the other members of the squad left in our room. Schumacher (you know him) greatly pleases me by his optimism. In spite of his years he is very strong and cheerful. His fate is very tragic. In May 1941 he left with the Odessa theatre on a tour to Minsk (he is a machinist who made the stage scenery). However, during the first days of the war, when Minsk burned as a result of the bombardment with incendiary bombs, he ran 40 km on foot with his wife and some other artists from the theatre to the nearest station. His daughter stayed in Odessa — she's a student in the medical school. He thinks that his daughter was trapped by the German encirclement of the city, as, until now, there has been no letter from her at her address in Stalingrad. Schumacher is very talkative and often speaks about the theatre. He has worked in the Odessa theatre for 30 years. He is friendly with many artists of drama, opera and the circus. He tells us many stories and describes much of the customs of the theatre in 'peacetime', i.e. before the October Revolution.

Now after the departure of the 'specialists' from our room we have moved around: Veyberg sleeps above me, Schumacher under me, and I sleep in the same old place as before.

Monday 30th March 1942

Yesterday was Sunday, a free day in our department. I used this Sunday very fruitfully and wrote an article for the second card index of CITEI (Central Institute of Technical and Economic Information).[12]

The snowstorm continued to rage for the entire day yesterday and it was very cold, so that I was pleased to sit in a warm room and deal with my writing. True, the writing of the article did not progress very rapidly. On the one hand, the situation was not greatly conducive to concentrated thought (45 people in the same room). On the other hand, it was difficult to switch back to my old style of thinking. I jokingly told my comrades that now it would be easier to write an article

[12]AGS did not remember what that article was; could it be a translation of an article from German?

on the theme 'as I worked out 200% of a standard for the unloading of a pile of coal, or during the clearing of snow from in front of the brick kilns'.

During the evening of the 29th, our section was urgently sent to work on the unloading of railroad cars full of fine coke and coal. I was the person on duty after our bread supper, and so avoided this trial. More precisely, as morning approached, I went to help the brigade, and from 4.30 a.m. to 5.30 a.m. we finished unloading the coal from the last railroad car. Again we exceeded the production rate, which means that we will get 900 g of bread and a third cauldron: pea skilly with a fish in the morning, pea sauce (which is intermediate between a soup and porridge), a patty with fish during the day (30th March), and a skilly with fish in the evening. It is possible for us to survive on this if we continue to receive the same hereafter. The only problem is that the people in our section are not ready for such physical work both in terms of their health and their clothing: Rotermel (a fourth-year student of UII) has completely torn his felt boots; Alboni has no gloves; Veybert's got frostbite on his feet and he's physically weak. I am considered as one of strongest and the most appropriately dressed loaders in the section. My farmer's mittens, boots, Oleg's jacket, the quilted jacket — all of this perfectly suits the nature of my work.

By the way, Lidusya, it turned out that parcels to soldiers are not permitted. Only the criminals are allowed to receive parcels. It is possible that they will allow those soldiers who do not have footwear to go into Sverdlovsk city in the next two to three days, so that they may go to other soldiers' apartments and collect the necessary things for them. In this case, Lidusya, please collect and store a number of objects in one place at home and tell Granny Olga and others about this too, so that if someone from our section comes with a note from me then this package can be given to them even in your absence. If possible can you can put the following in the package: two tubes of calcex, Guggenheim's *Modern Thermodynamics*, a general notebook, a bottle of glue (for envelopes), paper, envelopes, postcards, stamps, onion, a little sugar, etc. I want to try to keep myself occupied. Ask Modestus. Maybe he can lend his book *Articles on the Physics of the Atom* temporarily to me via a messenger. Try (as a test) to send any fiction to me by registered parcel (for example, Chekhov's volume of short stories). If the book reaches me, then other books can be sent afterwards.

I have got a leave warrant to go into the city to send parcels to the UBAS.[13] I will send this letter to you from the city. Write to me saying what letters you have received from me, with dates.

Your Armin.

Letter 7[14]

Tuesday 31st March 1942

Dear wife!

I have worked in my new position for 11 days now. I'm healthy, cheerful — that is most important. Starting from tomorrow I have obtained a promotion — I will be a dispatcher. I have written six letters to you; and have received not one from you so far. Write to me with your address stated on the back. Yesterday I saw Maria Samsonovna and got on agreeably with her. I wrote and sent the article to CITEI by registered wrapper. Has Al. Iv. got the money for me? Call Murzin. At the end of April phone CITEI. Also, ask them if UBAS sent the second article to them. Write about our daughter. Do you have firewood? Does she attend kindergarten? Send Guggenheim's *Modern Thermodynamics* to me by parcel to the same address. Send me the addresses of Modestus and Tosha.[15] Regards to all our loved ones. Did Adrian get my letter? My thoughts and my feelings are together with you.

Your loving husband.

I sent this postcard 2/IV together with the letter No. 9 through the mailbox on our street.

---

[13]Ural Branch of Academy of Sciences (UFAN, *rus. УФАН*), in Yekaterinburg.

[14]A postcard. The address on the postcard: N. Tagil, Ural St 3, Ananieva M.S., to me — EAZ.

[15]The nickname of Antonina, Lydia's sister.

Letter 8

Tuesday 31st March 1942

My dear Lidusya!

I begin to worry about the lack of any letters from you. Yesterday I obtained a leave warrant for the city and travelled around the city from 4.00 p.m. to 7.00 p.m. I visited the post office, and sent a registered parcel to the UBAS with the second article in CITEI. Truly this was a heroic act, taking into account the situation of writing this article. I sent a telegram to you. After this, I went to Maria Samsonovna Ananieva. She lives in a comfortable room (with two children: a daughter of twelve years old and a son of ten) in their own house almost in the centre of Nizhni Tagil. It's approximately one hour (4–5 km) on foot from my barracks to the centre of Nizhni Tagil.

Today, once again, our brigade cleared snow from the area in front of the furnaces. The number of people in our section is melting away. Twelve of the original 36 men are now left in our section (eleventh brigade of second column), and only five people worked today (without taking into account our brigade leader, Remshevich). Namely: Schumacher (50 years old, the scene machinist), Veybert (19 years old, the first-year student of the UII), Trelin (a shoemaker), Lots (a tractor operator) and I. The other people who were out of work from the brigade: Kirshteyn (an electrical engineer) — a task master; Ulrich (an electro-welder) — stays at home because he lacks any footwear (he burnt his felt boots on the furnace); Tribelgorn (a designer of coking) — an estimator; Alboni (a mechanical engineer) — he stayed behind as the person on duty after the bread supper; Ryukkert fell ill (with a cold), Rotermel (a fourth-year student of UII studying mathematics) — he fell ill (a cold). Our brigade of loaders is increasingly becomes less effective. Rotermel's felt boots were torn completely (he only has tarpaulin shoes); Alboni has no mittens; Veybert is physically weak. Thus far Schumacher is still stronger than the others (without considering Tremin, Lots and me), although he complains constantly about his liver. We still exceed the normal standards, since our task consists of a large number of small commissions. But if our brigade is sent to work full-time (for example, unloading bricks from the furnace), then surely the brigade will not carry out the standard rate.

From tomorrow I obtained a promotion: I am assigned as a dispatcher in the shipping shop instead of Shtol (a mechanical engineer from our party), who will become a mechanic at the Hare Mountain site (the clay quarry). Taking into account the aforementioned state of our brigade, I decided to agree to hold the post of dispatcher, although there is nothing enviable in it. Suffice it to say that a semi-literate girl is one of the dispatchers in the shipping shop. I will regularly get 700 g of bread for 12 hours work and then 24 hours rest (day and night shifts from 6.00 a.m. and 6.00 p.m.); I will pass into the special column and I will live in the ITS room.[16] Tomorrow morning I begin to work in this post. So far, I do not see any other prospects of a job.

Letter 9

Wednesday 1st April 1942

Dear Lidusya!

It is now already April; then there will be May, June... So this sad period in our lives will gradually pass. *April, April! Er weist nicht was er will!*[17] Today a sharp, piercing wind blew for the entire day, and the snow melted on the leeward side.

Today was the first day that I worked in my new post — as a dispatcher in the shipping shop, or simply in the post of a foreman; or even more simply as the 'hurry up' man. I want to share my first impressions with you. These impressions are far from enthusiastic, but nevertheless it's better than coaling work. My job concerns the fact that it is necessary to ensure the smooth delivery of fuel, brick halves (broken bricks) and slag to the brick plant; to ensure that the driveway and loading areas are cleared. The manpower consists of two brigades of loaders — Germans (including one brigade from Sverdlovsk: Ryukkert, Schumacher, etc). Today supplying the 'crusher' with brick halves was a bottleneck. Peltser was the shift's dispatcher (formerly he worked as a guard at the railway

[16]ITS — engineering and technical staff — ASK.

[17]From the poem of Heinrich Seidel — ASK.

station), and I walked with him to be 'trained'. We had to find the deposits of the broken bricks under the snow, to assign a brigade to dig these bricks out; to assign a part of the brigade to clear the way to the truck; to instruct the driver; to find crowbars and irons in the smithy, etc. Thus during the day each of us ran approximately 10–20 km over the entire area of the plant… But nevertheless it is better than rummaging with a pick amongst a heap of broken bricks. At the end of the job we had to fill in the work orders for the brigade leaders. Poor Peltser! He was drenched ten times over before he had calculated the cubic meters of bricks that had been dug up by the brigade. Then, when we went to the plant, Peltser, sighing heavily, frankly acknowledged to me that the writing of the work orders is the height of wisdom for him. Tomorrow in the morning I will again double-up with him, and then the day after tomorrow, hopefully, I will work independently. Tomorrow they will move me, and I will become employed in the staff of the construction crew.

That's the matter, dearest Lidusya. Still there has been no letter from you, but I hope that over several days I will obtain news about you from someone coming from Sverdlovsk. We will see what is going to happen next? We have survived ten days, and maybe we will survive for longer. Write to me at the address: N. Tagil, Ural St., 3, to Ananieva M.S., for me.

Thursday 2nd April 1942

At long last yesterday, in the evening, I received a postcard dated the 27th of March from you, dearest Lidusya. You do not mention that you have received any of my letters, how many and when. You must write about it, and also number your letters so that I will know how well they reach me. Does our daughter order me to come back home sooner? Tell her that I will arrive on the same day that they release me from the army. If Rudnitsky, Simonenko and Obukhov get the reservation, then probably Rudnitsky can be sent back to the UBAS. Does UBAS concern themselves about me? On the 28th of March I wrote a postcard to Chufarov and to Adrian [Smirnov — EAZ], on the 30th, a letter to P.A. Volkov. I asked Adrian and Volkov to make enquiries about this matter, but as of now I have not received any answers to these letters. Could you, Lidusya, visit the UBAS and ask Chufarov or Novitskiy about what the UBAS has

I have described my location in detail, since your arrival seems to me to be increasingly likely after yesterday's conversation with Bernhard, than it seemed before. It would be good if you could find a few other places in Nizhni Tagil so that it would be easier to find lodgings for the night and leave your belongings; maybe ask Bernhard's wife or even one of the neighbours. Subsequently, maybe they will permit me to live with my wife in a private apartment (I'll ask about it in detail). We will rent a room in one of the houses on the adjacent streets and I will live for a week together with you. Dreams, dreams…

By the way, Lidusya, send me the address of Nina Kroll (tel. 1117), Modestus Kler (tel. 6529 and home 7859), Elga (ask mum), Vladimir Krichagin (also ask mum), Antonina Mikhaylovna Poponina (Is it Bolshakov Street, 76?).

I received your postcard on the 2nd of April (written on the 27th of March) — the first postcard from Sverdlovsk; and on the 3rd of April a postcard from Adrian. He writes that a reservation has been obtained for me. I wrote a letter to him today with a request to visit Novitskiy and to ask what the UBAS has undertaken. Please phone Adrian, ask him whether he carried out my request and write to me about this. By the way, send me the telephone number of Shlygin.[19] I will not disturb them unnecessarily, but if there is something urgent, for example, to arrange the details of your arrival, then I could phone you there. Bernhard also spoke to his wife's sister using one of his neighbours' telephones. I await a letter from you with a description of all the family news, the UBAS and the UII.

Sunday 5th April 1942

Lidusya!

Today I spoke with two men who arrived here from Sverdlovsk. The first of them, Ulrich, is an electro-welder from our section. He made a semi-official trip to Sverdlovsk (without a ticket to the city, and back with a ticket) to get some footwear and he tells us that he will not go back there ever again (his wife is another matter). I asked him to phone Modestus so

[19]Their next-door neighbours at the time — EAZ.

that you could drop in with his address. But he did not do this. (He says that no one answered the phone.) Maybe he was afraid that you would bring too many things for me. In the middle of April his wife Tatiana Petrovna Morgunova will arrive. Her address: 3rd kilometre of UZTM (Ural Heavy Machinery Plant), the house of the mechanization No. 2, Ulrich's apartment. Please send some things with her, not too much so that it would not be cumbersome or heavy for her, for example: 20 onions, a general notebook, some paper, envelopes, 200 g of sugar.

The wife of our soldier, Taborsky, arrived together with Ulrich. In the case of your arrival (in the more distant future) I inform you that it is necessary to be supplied with papers from your husband's former workplace (as Taborsky's wife had) of approximately this content:

Certificate

Given the fact that the former senior scientific worker of the Ural Branch of the Academy of Sciences of the USSR Armin Genrikhovich Stromberg was mobilized on the 16th of March 1942 in the RKKA,[20] this information is given to his wife Lydia Mikhaylovna Stromberg in support of her passage to her husband Armin Genrikhovich Stromberg, who is located in the construction crew in Nizhni Tagil.

Stamp Substitute presidium of the UBAS / Demenev/
Head of the secretariat....

Taborsky's wife purchased her ticket along with this certificate and presented it to militiaman in the train (her return ticket also was purchased using the certificate shown).

The second visitor from Sverdlovsk was Bernhard's wife, Helen Grigorevna. She is very concerned by the fact that she did not tell you about her departure. This may be explained by the hurry in which she left, since she arrived together with the wife of our soldier Scheurer. Helen Grigorevna got a mission certification (fictitious) stating that she will go into the kolkhozes [collective farms] of Nizhni Tagil region in connection with an application for some seed funding for a project.

---

[20]Workers' and Peasants' Red Army.

She has arranged to live two blocks from our barracks (Red Flag Street, 62, second door in the court), where she will live in a separate comfortable room together with Bernhard for several days.

Monday 6th April 1942

Today I'll be on duty from 6.00 a.m. to 6.00 p.m. I came home for breakfast and decided to write you a few words. Bernhard's wife will leave on the 8th or 9th of April together with one of the staff workers. Until this time I have decided to gradually transfer all my letters to her, so that you will receive the whole bundle.

I received two postcards from you yesterday (the 5th of April) in the evening, stamped on the 30th of March and the 31st of March (the first was written on the 27th of March) and another postcard from our daughter. The drawings are very good, but it's difficult to guess the sense of them. It seems to be an aircraft, tank and a gun. The signature 'Elya' is remarkable. I'm very distressed by our daughter's illness. I hope that she gets well soon. You write that a plea for exoneration has been obtained for me and that the UBAS is getting on with it. I want to believe that this will lead to some positive results. Lidusya! Please visit Novitskiy and Chufarov. Firmly enquire as to what is being undertaken and write to me. If the UBAS wants to write to Nizhni Tagil, probably the best address is: Nizhni Tagil, *Tagilstroy* NKVD, construction crew 18–74, to the commissioner or the commander. However, it is not entirely clear to me why it is necessary to write to Nizhni Tagil? Thus far I have heard nothing about any case of demobilization for anyone from the construction crew. I personally pin my faith on the course of military events. If Hitler were to be smashed, then they will demobilize us soon.

Yesterday, in the evening, I went to the apartment of the couples of Bernhard and Scheurer (they rent a room in a private house not far from the barracks…). Happy wives entertained their happy husbands with tea from real cups and with various rolls. They began to entertain me, but I felt myself to be surplus to requirements and left. I promised to visit them later but did not.

I sent postcards (congratulations to Nina Croll with the 6th of April, and to the Lyalya station — to the relatives of Nikolay Leontiev), and then

I walked alone for approximately one hour through the melting snow on Red Flag Street… I was sad and dreamed. About what? That you had arrived and we were the people sitting in this room; or that I left and I am with you, together, in Sverdlovsk again.

Tuesday 7th April 1942, 8.00 a.m.

Lidusya, yesterday I received the letter from you dated the 2nd of April 1942, the fifth correspondence on the list… Thus I have received all of your letters (my letter didn't pass the military censorship). I hope that now our correspondence will be restored. You, Lidusenka, must write to me more often, since your letters are a great joy in my forced exile.

I am amazed by the literacy of our daughter. I simply do not believe my eyes when reading the words written by her hand.

You ask whether I require any money. I can report to you that until now money is completely superfluous to me; I have bought nothing besides stamps for the letters, and there is nothing else to buy. I have enough bread, frequently obtaining between 700 g (100% standard) to 900 g (which is more than 110%). It is rumoured that it is possible to exchange milk for bread: 1 litre of milk for half a loaf. I want to try to economize on bread (for example substituting it with dried crusts) in order to barter half for 1 litre of milk from time to time. The problem of laundry will appear soon. But I think that it will be possible to find some woman to supply a piece of soap.

Since I became a dispatcher I'm not occupied with hard physical work, and our nourishment has also improved (they began to feed us for the third time in the daytime — lunch). I generally have no feelings of hunger. Only the monotonous menu annoys me: pea soup and herring in the morning, in the daytime and in the evening. The last time, however, they began to diversify the menu, giving us some noodles or some thick oat soup.

I will relay to you, as an example, the typical menu we had yesterday: in the morning — pea soup (too much water, too little pea) and half a herring (50–80 g); in the daytime — pea porridge (two wooden spoons), pearl-barley porridge (one wooden spoon), herring (30–50 g) and a cup of tea with sugar. Some joker said that the sugar only lay next to the pot — not in it. In the evening — cabbage soup from the left-over pearl-barley.

And of course there is absolutely no hint of oil, fat or meat in any of these dishes.

It is difficult for you, Lidusya, to be alone. I would share your labour in teaching our daughter with pleasure, but there is no way to do so. Now she's of an interesting age. It's a pity that in my daily routine I paid little attention to her during the recent months before my departure. Now I frequently imagine my daughter resting on her elbows as she writes words and sketches the tanks, which you send to me in your letters. Dear, dear daughter. Whether I will see her soon? Ask Modestus to print Elza's photo for me as a remembrance. I wrote to him asking about this already, but I did not receive an answer to the postcard. Did you say that the boil burst on Elza's neck? Poor daughter! This bad hereditary trait was inherited from you, Lidusya. Now, hopefully, she will soon get better. I hope that she will be healthy. She's been unlucky during this winter. Lidusya, feed her PROPERLY. Do not even think to send butter to me. There is some mention that it is possible to purchase 1kg of butter for 400 roubles. Sell my bright suit, but do not leave my daughter without any butter.

Today or tomorrow Helen Grigorievna will go back to Sverdlovsk. (Today.) I will again carry a bundle of sheets of letters to her.

Today I work overnight from 6.00 p.m. to 6.00 a.m. In the daytime I will try to arrange my business: to obtain an extraction order concerning my enrolment as a dispatcher and to be transferred into the ITS room of the special column. Although there are less people (two-storey bunks), they are cleaner.

Tomorrow I will try to go to the city. I will visit Maria Samsonovna and let her know about the parcel, about storing any belongings and any overnight stay (if you arrive). Maybe she will give me some book of fiction to read.

Wednesday 8th April 1942 1.30 a.m.

I sit at the dispatcher's table during my turn on the nightshift from 6.00 p.m. to 6.00 a.m. Another four and a half hours remain until the end of my duty. These are the calmest hours. Two workers finish loading the last railroad car with bricks; the truck will make two or three additional voyages to deliver *polovnyak* (broken bricks) to the crusher. I recall the disappointments of yesterday. Yesterday in the daytime the commander of our eleventh section suddenly reported that the commander of the second column ordered me (the dispatcher), Kirshteyn (a contractor) and Tribelgorn (an engineer-estimator) to return to the brigade of loaders.[21] The matter is clear — no one asked for the permission of the commander of column, and his feelings were hurt. I think that the matter will be settled, since I have an arrangement with the assistant commander of the construction crew, but nevertheless it is offensive. I'm like a serf. I wanted to be a nobleman, and he reduced me back to the ranks. I recall the novel of Mamin-Sibiryak, *Gordeev Brothers*...

They promised to provide the extraction order tomorrow, about my enrolment in the post of dispatcher, and then my position will be more tenable. It is terrible to think that I might have to be a loader again. Our soldier-loaders are terribly exhausted. Only 36 out of the original 78 people who arrived together with me from Sverdlovsk remain in the brigade of loaders. Most of all I feel sorry for poor Schumacher. After continuous loading and unloading, combined with his liver disease, he can hardly drag his feet and cannot eat anything. Yet the doctor will not excuse him from duty, since he does not have a fever. Veybert was ill for 4–5 days, Rotermel for 3–4 days. The members of our brigade are not ready for physical work and therefore they are suffering terribly from our poor diet. Ryukkert is noticeably emaciated. He's a bag of bones. In Helen Grigorievna Bernhard's opinion, I appear even better than on the day of my departure. For my own part, I actually feel rather good since I ceased working on the loading.

Wednesday 8th April 1942 3.30 a.m.

My shift has come to an end. It was my first independent shift. It seems I managed this 'complex' problem well. I piled up coal in the machine for the kilns, ensured that the crusher was filled with broken bricks, and loaded the railroad cars with more bricks, and wrote this letter to you...

[21] i.e. that they be demoted from their new jobs — GGW.

Did you ever imagine me in the post of a foreman at an antediluvian brick plant, where our feet substitute a telephone? Well, it doesn't matter. It is even healthier than, for example, to be a contractor or a norm-setter and to sit all day long in a smoke-filled room. Instead, a foreman works outside for the whole day. From the garage to the crusher, from the crusher to the kilns, then off to the office… so the whole day you run round. But then after a 12 hours shift you have 24 hours on your own, and don't have to think about work. For me this is a completely unusual state, specifically, the large amount of free time, free from the interests of work. I would like to use it somehow. I even think about trying to study some thermodynamics or mathematics. At the same time as you send the book to the address UL Ural 3, please send a postcard about it to me at the post office 2, P.O. Box 67.

Tomorrow Helen Grigorievna Bernhard and Lydia Georgievna Scheurer are going to leave. Immediately after my shift ends at 6.30 a.m. I will carry these sheets of letters to Helen Grigorevna and will take leave of both wives… Lidusya, what do you think the chances are that you can visit me here in Nizhni Tagil, and for how long? This would surely be possible in spring or in June, since the students' examinations will have begun. You can leave our daughter for a few days in the care of Granny Olga (if our daughter will be attending kindergarten, then this is not so bad). Nevertheless, I firmly hope that during June we will see each other. Of course, it is still necessary to chase the UBAS about my return, but I am pinning my faith entirely on external events. The worse for Hitler, the better for me and vice versa. Kiss you, Lidusenka.

Your husband-foreman, Armin S., DPhil.

Yesterday, the 7th of April, I received a postcard from you dated the 2nd of April, in which you report about receiving my letter No. 5. But what about the remaining letters? How many days does it take to receive letters from me? It's four days from you.

Lidusya! If you decide to come to Nizhni Tagil as a result of a conversation with Bernhard's wife, first bring along the boots with the overshoes (due to my post as a dispatcher I am always splashing through puddles, thus walking with wet feet) and some dried crust (I still haven't touched the dried crust you sent earlier, so that is not compulsory). It would be good if you could bring some other books:

Blackwood	*An Outline of Atomic Physics* (Modestus has this book, ask him for me, he will probably let me have it).
Smirnov	*Higher Mathematics, Volume I.*

Do not forget to bring some paper, envelopes, glue and an adversaria,[22] since it is prohibited to receive a scientific paper in a parcel.

Letter 12

[Letter 11 did not survive.]

Wednesday 15th April 1942 19:00

Dear wife!

On the 13th of April Bernhard gave me your parcel and letter. I thank you very much for the fulfilment of my request. Now I have literally everything what a man can wish for in my situation, namely: clothing, nourishment and implements for education and the writing of letters. How often I have mentally thanked you a thousand times over for all the trifling things that would seem unimportant, but that you lovingly assumed I would need and packed into my bag. My farmer's boots, until now, have coped perfectly well with the mud of Nizhni Tagil, so, thus far, do not worry about the boots with the overshoes, just bring them with you when you come to visit me. Coming home, I throw down my farmer's boots, put on the dry socks and the tennis shoes and feel as if I were at home. I was recently occupied with the repair of drawers and trousers — everything that was necessary for this (patch, thread, needle, scissors) was available on tap. Everything that is required in order to write letters is also provided (inks, glue, quill, paper). I am gradually eating up the cheese and brisket as a delicacy. Thank you so much for the sugar and the onions. By the

[22] A peer-review report on a scientific paper — GGW.

way, I also managed to obtain some onion. I wanted to barter some milk in exchange for some bread, but instead of this they exchanged a full plate of onions for 400 g of bread. The addition of onion into the skilly along with some black bread greatly livens up the process of eating. However, our nourishment has continued to improve. For example, yesterday in the evening they gave us oat porridge (quite a big amount) and half of a fat and juicy herring. I even have some bread left, thus I have collected about 1 kg. I will exchange it for some milk. Food produce cannot be obtained for money. Someone told me that a loaf of bread on the market costs 100 roubles, whilst a litre of milk costs around 30–40 roubles; the price of potato is a meaningless phrase, since it does not appear on the market.

Nevertheless I have spent a little money. On one occasion I purchased some special gloves — a leather palm and thumb and with fabric on the back side — for 30 roubles from one soldier (a brigade leader that I wrote an order for). Now it will be possible to save the fur gloves until autumn.

I spent about 30 roubles on books.[23] I purchased four books in the Nizhni Tagil office of the Book-trading Union of Government Publishing Houses: Pavlov — *A Brief Course of Organic Chemistry;* Pavlov and Semenchenko — *Inorganic Chemistry;* Alabyshev and others — *Electrochemistry of Fused Salts;* Dubinin — *Basic Physical Chemistry of Sorption Processes.* This is all that I found interesting in the shop. Now I have begun to study Pavlov's book with some enthusiasm. It is so pleasant to come home after a tiresome and uninteresting day as a dispatcher, scurrying around the plant, to sit down in a warm corner of the room and to dive into the world of organic compounds. I have never studied with such pleasure. Impatiently I await the parcel with the Guggenheim book.

I still live in the same room and at the same place. Schumacher is beneath me, and Veybert is above. Only 12 of 38 people are left in the room (and the section) of those who arrived with me from Sverdlovsk, and nine new people were added, so that the section is now 21 people. Fourteen people work in the brigade of loaders in the delivery shop, i.e. under my direct management, and seven people from the section have

[23]As a dispatcher, I had a twenty-four hour pass through the gate of the camp — AGS.

another job. These loaders work in two shifts: seven men in two shifts (4.00 p.m. to 12.00 a.m.) and a second shift (12.00 a.m. to 7.00 a.m.). Every third day I'm on the night shift (7.00 p.m. to 7.00 a.m.) and then I write orders for the brigades. Kirshteyn, from our section, is a norm-setter who writes the standards for the work rate. The brigade's work is not too difficult: collecting coal into heaps, levelling the area, picking up broken bricks. The work with the bricks is the most unpleasant: unloading trolleys (from the furnaces) and piling bricks, loading bricks onto the flatcars. The main problem is that they do not give us any gloves and there are not enough aprons. All members of the brigade have torn gloves, their hands are covered with bleeding wounds, and their clothing is ripped and torn. That's the main reason I am happy to move to the dispatcher post. But the brigade exceeds the standards regularly (this is also a credit to the brigade leader, dispatcher and norm-setter) and my comrades get 900g of bread and a third cauldron, so that they do not starve.

Generally, there is little pleasure in the dispatcher's work. My boss — the chief dispatcher — is from among the special migrants with little education who rose to that post — a typical foreman. He is a foul-mouthed man, he fusses and meddles in everything; he shouts at me like I'm a boy (allegedly, a German dependant). But I'm not offended because actually Ivan Arkhipovich [Loban — AGS] is not so bad as a person. Apparently, the shouting and obscenities are a conventional system at similar, primitive, provincial plants.

Thursday 16th April 1942

Recently I read in 'Nizhni Tagilsk worker' dated the 14th of April about the presentation of Stalin awards. I'm very glad that there are so many from the UBAS staff among them: Kikoin,[24] Gubar, Obukhov, Mikhaylov, Sigov, Gaydukov. There was no one listed from the UBAS in the past year. This is, of course, connected with the wartime conditions: with the fact that the centre of gravity of industry moved from the Urals; with the fact that defence-applied themes get special importance. It is typical that no one from the UBAS received any awards for their scientific work.

[24]Grigoriy Constantinovich, from the Leningrad group — AGS.

It is pleasant to see many great scientific names among them: Fersman, Ioffe, Zelinsky, Rehbinder, Mandelstam (is he your teacher from Leningrad State University?).[25]

It is pleasing that the first three awards, as an exception, were offered to mathematicians and physicists. Upon reading this decision, the project titles and scientific names, I was reminded of my scientific work and became sad to be in this forced inaction. Well, not to worry. I will try to fill this gap by working on a book.

Time goes by. Today is exactly one month since the day of my mobilization. I remember as it was exactly a month ago, on the 16th of March, that I rushed into the UBAS, took my leave, completed deals; meanwhile there was a devilish spirit among the tailors at home. Everyone sat and sewed: you, Granny Olga, Mum, Lelya — I changed clothes on the run, grabbed my belongings, and ran to the deportation point… Then there were three wearisome days of waiting to depart… These were, perhaps, the most difficult days for me. A sort of uncertainty. A hope that everything was only a bad dream. Now it is very clear that this is reality. But habit is second nature. If I could only see you and our daughter, even just once. I want to believe that you will visit in May at least. It would be such happiness! Well thus far, Lidusenka, don't worry about me. Do not forget to write me. N. Tagil, Post office 2, P. Box 67, column 2, section 11.

Your husband.

Letter 13

Tuesday 21st April 1942, 5.00 p.m.

Dear wife!

These past few days (beginning on Sunday) were troublesome. It was decided to reorganize the work force according to the type of work required and, correspondingly, to reorganize the sections; for example, the section of loaders of the shipping shop (which I was a member of

---

[25]Kapitsa is not in the list, since later he refused to participate in the atomic project, therefore he lost the post of head of institute — AGS, 2003.

initially), the department of mechanics, etc. I was transferred into the seventeenth section of the third column — ITS section, consisting of storekeepers, timekeepers, work superintendents, etc., but which also includes some real engineers as well. The works manager is now called 'the chief of the special camp of the NKVD *Tagilstroy*' (according to the inscription on the door of his office). The commander of our detachment is now called 'the assistant-chief of the special camp of the NKVD *Tagilstroy*'. Until now, I did not understand the actual meaning of this terminology. Everyone has the feeling that some change for the better may be taking place. One of the soldiers returned from a mission in Serov city and said that the construction crew of Germans in Serov city had been dismissed (I heard this from another soldier who had spoken to this particular man, so the authenticity of this information cannot be been guaranteed). They demobilized the aged, sick, disabled and some people who were deemed to be of no use, whilst the requisite number of specialists and manpower remained mobilized in the enterprise, but as civilian workers: they live in a private room, earn a wage and food ration ticket, are allowed to buy food for money in the canteen; they enjoy complete freedom at the end of the working day. Thus far we know nothing about it in here, so that it is perhaps not entirely true. Nevertheless, such rumours give us some confidence that the changes taking place may lead to a better situation for us. In the meantime, they are surrounding our barracks with fences, probably with barbed wire and a control post.

Today (since Sunday) I have been moved to a third room and, again, it is only temporary. At present I am arranged sufficiently comfortably: the second floor of the same tenth barrack. It is a room for 'diverse' people, in the event that there is not enough space for them in the room of the corresponding section. I've got the lower cot (here there are only two-storey bunks), isolated as in the sleeping compartment of a railroad car, near the window to the west. Now the sun shines into the window directly onto my face. Summer comes soon, it will be hot and dry … and events at the front will begin. I wish that Hitler would be defeated faster by the united efforts of our troops, England and the USA. I wish that peace be made soon so that I could return to the UBAS and be occupied with my labour of love!

But thus far I have to accept the situation. I am a dispatcher as before, but I do not know how long I will hold this post. On this very day they discharged one German dispatcher for the fact that during the loading of the bricks he missed some and sent a half-loaded platform. Well, there could be the same incidence of my absent-mindedness too! In any case, I try to display quickness during the discharge of my duties, so that I have a good relationship with the authorities and the local dispatcher. Today, at night, we worked in shift together with one of them — Peter Afanasievich Kazimirov — and sitting in the office we spoke heart to heart. He frankly gave his opinion about me: 'You manage the work well, but … be more energetic with the workers, swear at them, then they will feel more responsible and will work better!' Incidentally, timid 'expletives' begin to escape my lips when dealing with the workers, and it seems that they make my orders more accessible. He [Kazimirov — GGW] lives at the brick plant without a passport in the position of 'special migrant', although he assures me that he is not. The loading shop chief Ivan Arkhipovich Loban (my direct superior) is a special migrant without a passport too. Many of the staff — men and women — are also from the migrants (dispossessed kulaks and others).

Gradually I am mastering the work. I have performed a few critical tasks already. For example, I travelled to the commodity station 'Western' in search of a railroad car with coal for the brick plant. At night without any documents I walked through the entire Nizhni Tagil, climbed on the mounds, crawled under the railroad cars, walked through mud and puddles, until finally I reached my goal, and after visiting 45 people (the person on duty, the filing clerk, the dispatcher, etc.) I finally found this ill-fated van; now it has arrived and has already been unloaded. Another time, a locomotive arrived unexpectedly in the area of the plant to collect 14 empty wood-board platforms and a wagon of lime, and I sent them successfully, and even wrote the 'waybill'. But I did not know the destination address or designation, so after the shift I had to journey to the 'Limestones' station, which is about 1.52 km from the plant. Today two railroad cars again arrived at night unexpectedly. Loads always arrive unexpectedly in any given period of 24 hours, since it depends on the time when the locomotive is available to deliver railroad cars to the station 'Limestones' or 'Western'. These two wagons delivered 50 pigs for the subsidiary plot of the plant, and I had to look for the chief of supply at night in the barracks, so that he could urgently organize the 'unloading' of the pigs. Then I travelled at night to the barracks for a second time (about 0.5 km from the plant) in order to call the brigade of loaders (Estler's brigade) to load these railroad cars with bricks. Then it was necessary to give out tools (shovels) to the workers, to organize the unloading, and to distribute the workers accordingly. At the end of the shift I accepted some bricks from the kilning shop. The piles of 200 bricks were counted and whitewashed, slapping the piles of bricks with a rag at the end of some heavy-gauge wire dipped in a bucket of lime milk. Here, my time on-duty passes in the fulfilment of such and similar obligations. But then during the period of my day-long breaks I take great delight in reading Pavlov's *Brief Course of Organic Chemistry*, in writing letters, in repairing my clothing, and simply being at leisure.

Recently I spoke to Serge Vonsovsky.[26] He actually lives not so far from me, beyond the Tagil pond (about half an hour on foot). On the third visit I found him at home sleeping at 6.00 a.m. He told me a few bits of news. I want to visit him again on Thursday, the day after tomorrow. Schur should arrive by then, maybe he will bring some news from you.

I have drunk milk twice already. Half a litre of cow's milk from a certain Alexandra Polagaeevna in exchange for 350 g of bread. I drank it immediately after purchasing it. She has eight brothers, two of them were killed in the Finnish war, five of them are now at the front, and recently they sent the sixth [to the front — GGW]. Her husband is a turner at a defence plant; he is a reservist and they will not mobilize him for now. She's a very likable and talkative woman, a laundress; she lives not far from me, and I will obtain milk from her when some bread becomes available. Another time I drank some goat's milk for 350 g of bread + five roubles from Anisya Semenovna. I returned some linen to be washed: four pieces (jacket, drawers, towel and vest) for ten roubles (and without any soap!) so that there are no difficulties with the laundry yet.

I read in the newspaper about our Stalin laureates … and wrote congratulatory postcards to Kikoin and Gubar.

Here it is, Lidusenka! That's how I am slowly living. Today the second month of my stay in Nizhni Tagil began (I arrived on the 20th of March 1942)…, then the third will begin … and gradually the day will arrive when this sad period of my life ends. Write to me more often, Lidusenka.

---

[26]A physicist friend, the future president of the Ural Branch of Academy of Sciences — EAZ.

It's been ten days without any letters from you. Please write at least once every five days. Your letters — they are a big relief to me. I have re-read your last letter, dated the 11th of April probably more than ten times.

Your husband-foreman, A. Stromberg

Lidusya, how are your finances? Did you get the money on all my warrants?

(1) A.I. Zelyanskaya must obtain 100 + 300 = 400 roubles for me from the workshop at the UBAS for the calibration of the third polarograph and the plans for the construction of the new type of polarograph. Did she obtain this money? If not, then why not? I wrote a letter to A.I. in which I attached a letter to Goldobin, the manager of the UBAS workshops.

(2) CITEI must pay 400, 500 roubles to me for the two articles for the card index. Did you get this money? ... Call the chief engineer Sharfenshteyn (I gave you his number before) or visit CITEI with the address: Business house (on the 8th March Street), II floor, room 24 after the large hall. I sent the second pamphlet by registered parcel from Nizhni Tagil to Chufarov in the UBAS. Ask A.I. Zelyanskaya, or G.I. Chufarov, or O.I. Berezovskaya if they received this parcel? Is it re-printed? Has it been sent to CITEI with the attached letter to Sharfenshteyn and the information from UBAS about the introduction of the method in the Sverdlovsk Electromechanical Enterprise?

(3) From the translation bureaus of Nikolai Grigorievich Murzin for the translations:

    (a) 50 roubles for the partial translation of an article from German, approximately half a printed sheet;
    (b) 2030 roubles for the editing of a translation from German. Ring Murzin on tel. 7843 and ask him whether or not he made the calculations. I phoned him before my departure and he promised to pay. Lidusya, please write to me about all these monetary affairs.

Your A.S.

Letter 14

Saturday 25th April 1942, 1.00 p.m.

My dear wife!

The sun shines; it has become dry, and a light, a warm breeze is blowing. I am sitting on a log behind the calcining shop after breakfast and rest here until the workers begin to roll trolleys of brick after the midday break. Today I can hardly shuffle my feet and want to sleep so badly, as yesterday I became hyperactive, barely slept after the night shift and was astir the entire day. I decided to undertake an attempt (which is surely hopeless) to move to another job (to work with the polarograph) in the chemical laboratory of the 183rd plant. Until now, there was a slender hope via UBAS. But on Thursday the 23rd of April at 6.00 a.m. I talked with Serge Vonsovsky and Ian Schur; Schur said that before his departure he heard within the UBAS that the military registration and enlistment office had refused a reservation for Noskov and for me. Then Schur convinced me to attempt to be deployed in plant 183 to work as a specialist and to benefit the country's defence. I decided to make an attempt. I phoned the office of the chief metallurgist, Yudin, on the morning of the 24th of April. He advised me to visit the personnel office. During the daytime I phoned the manager of the plant's chemical laboratory, Zaikin, and arranged a meeting with him at the plant. After this I went to the locomotive plant to the personnel office (one hour on foot to the city centre and 40–50 minutes on the tram to the very end of the tenth route). But I spoke to the chief assistant of the personnel office; he said that it's almost impossible to come to an agreement with the current authorities. Then the manager of the chemical laboratory, Zaikin, came and we had a chat. He said that the chief of the central laboratory, Rodionov, has no objection to accepting me for employment. But there are difficulties with the formalities. I think that they are barely surmountable. By chance, Zaikin knows me after I gave a report to the house of technology at the conference of managers. On Monday I want to take a trip to the locomotive plant once again to clarify the situation.

Monday 27th April 7.00 a.m.

Yesterday I phoned Rodionov, and he promised to talk about me with the head of the personnel division of plant 183. I will ring him between 10.00 and 11.00 a.m.

I wanted to talk on Friday evening (the 24th of April) with you Lidusya, but had no luck. I purchased a receipt for this call with Sverdlovsk with my request and rang for an hour unsuccessfully to the long-distance phone station from the office of the brick plant. They finally said that the line was occupied and they requested that I ring back at 2.00 p.m. I decided to forego this pleasure and left to sleep. It is interesting — had they warned Modestus about the phone call?

This morning (the 27th of April) your letter was given to me, dated the 18th of April, with the attached letter from Elza. She gives you a lot of trouble. If only I could share these troubles with you, it would be such a pleasure! Elza is developing rapidly! Indeed when I left (the 19th of March) she could hardly write her name, and now she writes it even without copying the letters. Lidusya, tell me more about Elza. What she does say about the kindergarten, about the other children? Surely she's glad when you come to take her from the kindergarten. How do you manage to live without any electricity or kerosene? It is really good that Elza attends the kindergarten.

I received the book of Guggenheim's *Modern Thermodynamics* (I was in the city on the 23rd of April and called in at M.S. Ananieva's on my way). Thank you very much. After Pavlov I will start Guggenheim. Perhaps, do not send Blackwood's book for now. Try to send some inexpensive fiction (for example a volume of Chekhov's short stories). If that book reaches me then I will send it to you conversely after reading it and thus it will be possible to organize an exchange of books.

I now live in the room of the seventeenth section together with 29 engineers. Of course, it is much more pleasant to live in this room than in the previous one: the air is clean and we do not have these constant quarrels about sharing out the food. I sleep on the second shelf near the wall on a six-bunk, three-storied 'set of shelves' next to Fridrikhsen — another dispatcher from the loading shop. He was an instructor of physics and mathematics at a rural school in the North Caucasus. His great-grandfathers moved from Germany to the Crimea during the time of Catherine II. His father was a farmer in the Crimea. Initially Fridrikhsen evacuated to Kazakhstan with his wife, and he was mobilized from there. I have made no special friendships with any of the engineers, so that I do not feel myself to be completely comfortable in the new room. I feel sorry about one assistant of the organic department of Sverdlovsk State University — Shpadi. He was evacuated with his wife from Sverdlovsk into the Ivdelsky region;[27] he taught mathematics in a rural school; then he was mobilized to the Nizhni Tagil camp. He arrived somewhat earlier than me, but up until now he's worked as a loader. He's an extremely modest young fellow and very likable. I want to get to know him better.

We now eat in the dining room three times a day (at 6.00 a.m., midday and 6.00 p.m.). The entire department marches to the dining room in formation, with our own spoons and bread. Of course, this is more hygienic and we obtain bigger portions; but the meal takes more time and the process is not very organized yet. Our diet has been improving gradually. They have begun to give us some meat. Yesterday in the daytime there were cutlets consisting of 20% meat and 80% bread, but nevertheless this was real meat, and I ate it with great enjoyment. This morning we were provided with soup with dumplings and kidneys.

I read an order that the brick plant is now renamed the detached camp section of the NKVD *Tagilstroy*; the works manager is the chief of the camp section. All the civilian workers and the entire personnel of construction crew 18–74 are included in its staff. The commander of construction crew 18–74 is now designated the chief assistant of the detached camp section of the NKVD *Tagilstroy*. They feed us according to the standard allowances for a work camp or an NKVD colony. They fenced in our barracks (up until then only the fence-post holes had been dug). Time will tell what this means. In any case there is nothing to console us in all this.

Yesterday in the evening I phoned Lisa Chernobrovkina. My crewmate Wolf gave me her phone number. Once in 1935 or 1936 Wolf and

---

[27]North of the Sverdlovsk region — ASK.

Maria Sergeevna Ivanova invited Lisa and I to a family celebration. I had not seen Wolf since then, but I met Lisa again while teaching practical physical chemistry in the Ural Polytechnic Institute, where Lisa was a third-year student specializing in pyrogenics. After that I lost touch with her. She lived with her husband in Kharkov. Her husband is in the army, and she was evacuated with two children to her mother in Nizhni Tagil. She lives on the 57th Newspaper Street about two blocks from the centre towards the river (the centre is the final stop of all trams on Lenin street). Today I will have another discussion with her so that she might permit you to spend a night or two to store belongings with her if you arrive in the evening on the local train.

Do you think it will be possible for you to visit in May or not? What about your teaching in May? When do the examinations begin? Many wives visit with mission certifications.

Recently I received a postcard from Nina (and replied to her) and a letter from Modestus. Nina wrote that she is going to visit you and, probably, she will be packing up her things for a year or so. Modestus describes the episode of meeting you and Elza in the cafe and the store on the First-of-May Street. He walked hand in hand with Elza along the street on a sunny day. I want to walk with my daughter! Indeed she asks a lot of questions in her fourth year: this is the most interesting age. It is good for her to go to kindergarten. It will give her some social skills. At least she is not alone during the daytime. Modestus promised to send some photographs of Elza. Remind him about this as often as you can. I would very much like to look upon my daughter even though it is in a photograph.

Tell me, Lidusya, more about all our Sverdlovsk friends and family. Write about your life, about the general motion of life in the city. Do you have any potatoes? How much do potatoes cost? A litre of milk? You write that the dairywoman rejected your offer for milk. How have you managed to obtain any milk? They surely will only exchange milk for bread. Write to me, Lidusya. Letters and books are all I have in my life for now. The rest is somewhat external, random, secondary.

Your Armin.

Please give me the phone number of Shlygin.

Letter 15

Tuesday 28th April 1942, 6.40 a.m.

Sweetheart, Darling!

If only you could know how much your last postcard dated the 21st of April gladdened me, the one where you report your plans to visit to me in Nizhni Tagil. Unfortunately, the postcard had a long trip to get to me: the postmark is the 26th of April, and they gave it to me only yesterday (the 27th of April) in the evening just before bedtime, so that you will probably not receive my reply yet. I wanted to send a telegram to you, but it requires going into the city. In order to go into the city it is necessary to have a leave warrant from the commander of column. And in order to have the leave warrant it is necessary to have a very valid reason. But unfortunately I do not have such a reason. Furthermore, I do not even know who is responsible for the leave warrants now, since today they changed the commander of our third column (there is some talk that the new commander is more niggling and stricter). However, it's too risky to go into the city without a leave warrant. They warned us that before the 1st of May any leave warrant will not be given without a valid reason; and that the documents are now being checked frequently in the city. Yesterday at the evening roll call they also notified us that ten people from the construction crew were caught in the city without any leave warrant documents. Now they are imprisoned in the guardhouse, and they will be committed to a military tribunal court. Two of them were already condemned at an open session of the military tribunal (as a lesson to us all) and they sentenced them to eight years imprisonment and corrective labour for the fact that they were absent for 12 hours, namely they walked into Laya village searching for some potatoes (they left in the morning and returned in the evening, moreover they were absent from work). Absence in the construction crew for more than two hours without a valid reason is considered as desertion. Since I do not want to reside in the guardhouse when you arrive, then I must forego the thought of sending you a telegram. Of course, if the opportunity arises I will do so.

You ask what to bring for me. Without playing the hypocrite I can say — nothing. Certainly, it is not detrimental to have a coat (for the rainy weather) and shoes with the overshoes, a shirt and shorts for the very hot weather, fats = glycerides of stearic, palmitic and oleic acids, as I read yesterday in Pavlov's handbook *The Brief Course of Organic Chemistry*. Incidentally, I have worked through half of this book already.

However, it is the third day since they gradually started to give us some fats. For example, yesterday in the daytime there was pea porridge with several pieces of kidney. This morning there was a cabbage soup with dumplings, gristle and some fat, floating on the surface (the first such case in all the time of my stay here).

The weather changed from summer to autumnal yesterday, and today — into winter. It is cold outside, it is damp, it is dirty, and snow fell. The weather is the same as it was during October last year, when I was on the harvesting campaign in the Butkinsky region.

Nonetheless, today my mood is very good, as opposed to yesterday. Guess why? If only you would come, we can talk about our separated lives in detail. Indeed we have been apart only twice since we first met. Since December 1937, the first time was when you left to go to Leningrad for your graduate examinations and then in the summer of 1940, when you and Elza left to Krasnoufimsk a month earlier than me. Now, once again, we have lived apart for a month, but what a big difference between those two cases and this one. Then we knew that the separation was temporary, short term, and we did not worry about the present and the future. Now the future is gloomy with uncertainty; and it promises nothing good thus far. But I believe, Lidusya, that the time will come, when we can begin to build our lives together.

Now let's move to purely practical questions. Where will you spend the night (if you arrive in the evening) or leave your belongings? It's possible to do both with Maria Samsonovna Ananieva, she lives on the 3rd Ural Street. (From the tram stop in the centre it is necessary to travel on Lenin Street one additional block along the left-hand side — in the same direction as the tram went — until on the pavement there are steep stairs leading 8–10 steps downward; after this, bear to the left into the alley, and then again to the right — this will be Ural street, parallel to Lenin Street. The third house from the corner — this is No. 3, the house numbers are missing.) Maria Samsonovna has three children: Lydia, 19 years old, a third-year student at the medical technical school; Alexander, 17, a second-year student of the mining and metallurgical technical school; and Faina, a seventh-year student at secondary school. I have told M.S. about the possibility of your arrival already. I will send her a postcard today in order to inform her about your possible arrival.

If there is no one at M.S.'s home, then you can make do and leave your belongings in the apartment of Lisa Chernobrovkina (Levitskaya, according to the husband's last name), who lives on the 57th Newspaper Street; the telephone: home 854 and office 1034 (cooperative 'Red Food-industry Worker'). Lisa Ch. is a pyrogenic chemist; but now she works in an office. She was evacuated with two children from Kharkov to Nizhni Tagil to live with her mother. Previously, I was good friends with her (in 1934–1935); we even played piano duets together in four parts, so she will willingly provide shelter for you. I will ring her today to tell her about your possible arrival. Also you can call her from the station. I have never visited her apartment, but I saw the 57th Newspaper Street. Lisa indicated that her house is two blocks along it towards the river from the city centre.

Now a few words about my address. The best way to get to me is thus (a little different from what I first wrote to you): from the last tram stop in 'the centre' (on Lenin Street) go on one block to the river to Carl Marx Street and on it bear to the left and up the hill two blocks to By Komsomol Street (this street is abutted against a white two-storey house with a red signboard at the door: 'The board of dining rooms and restaurants'). Go along Komsomol Street downhill to the bridge itself, cross the river, and continue straight on (after the bridge the street is called Frunze Street). Keep going straight on and eventually you will reach my barracks. Frunze Street is abutted against the plant fences, between which there is a passage, after a little while you will pass by the banks of a pond and finally you will turn off onto Red Banner Street, on which you should continue. After 34 blocks the street will cross the railroad line and on the right-hand side you will see the brick plant, at which I have the pleasure to be a dispatcher. Then you will begin to rise into the hills. Up the hill you will see to the left a wooden watchtower [*pozharku*], and to the right — a red-brick chapel. On the left-hand side against it is situated barracks No. 10, in which I live on the second floor in room 19, the second door from the entrance into the corridor, to the right (seventeenth department of third undertake Guggenheim's *Modern Thermodynamics,* which Lidusya sent to me.

Of course, the work does not please me greatly in itself. My chief is a special migrant (dispossessed from the Ukraine), a promoted worker. His principle is: if you don't fuss (even for nothing) and don't swear, then it means you are negligent in your official duties. In particular, when I'm on duty in the daytime, he always hangs around the plant for a whole day, he berates me (but he also gives a dressing-down to the other dispatchers on the shift), he interferes all the time. At first I was offended with this treatment, but now I have come to the conclusion that this chief is not a bad person, but obviously this is a tradition at this plant and now my relation with Ivan Arkhipovich is perfect.

Formally, my responsibilities are many since 'the dispatcher is the owner of the plant'. The timely supply of fuel, raw materials, and the loading of bricks (so that there are no railroad cars stood idle) — all of this is the responsibility of the dispatcher. But actually the plant is now replete with fuel, there are several trucks, Ivan Arkhipovich constantly hangs about the plant (especially in the daytime) and he resolves any difficult questions, so that there have been no incidents thus far. But nevertheless it is necessary to be always on the alert in order not to receive a punishment for some oversight. I especially enjoy being on the night shifts, since they pass much more calmly.

Wednesday 6th May 1942

Yesterday, Mummy, I spent the entire day trying to arrange my passage to plant 183, but so far without any results. I visited the head of the personnel office of plant 183, and then visited the military commissar at the Nizhni Tagil City Military Registration and Enlistment Office, and then the director of our brick plant. The latter stated: there can be no discussion of transfers. But I do not lose hope for the dialectical motion of the course of the events ('everything is in a state of flux') and to the proverb: that there is no rule without an exception. It is necessary to obtain a request from plant 183. Yet I do not know how to make this happen.

Your interesting and meaningful letter gave me a reason for reflections and recollections.

I clearly visualize the family picture, when everyone sits in the evenings with the last flashes of light and reads my letters. It is awful that you do not have electricity, kerosene and firewood — you are like moles. Even I live better — we have electric lighting and firewood (we collected some junk from the plant).

Elga also has a terrible, senseless situation. She cannot even return to Vladivostok now. We will hope that the information about the death of her father and a certificate will help to solve this trouble.

One should store up firewood in the summer. It is very unpleasant to me that now I cannot support you with funds. They never pay money to anyone, so far, but even if they do, then surely it would be some small change. They indicate that my payment as a dispatcher = 350–450 roubles. It remains to be seen whether any sum will remain, by and large, after deductions for food, loans and taxes.

Mum, it worries me that you are not eating well. If it is possible, let Granny Olga eat in the dining room on 10 Chapaev Street, and Lisa can have a meal in the technical school, and she can ask to take some small amount of potato home with her. You can barter potato for just about everything possible (ask Nyura to give some of Constantine's belongings for this). Belongings will come with time, as you always taught me; but if you now exhaust yourself with hunger, your health cannot be restored in any way. Mother, please do not sacrifice yourself for others. Think more about yourself and your relatives; others (Lisa etc.) will survive perfectly well by themselves.

I wrote a postcard to Vassa Nikolaeva. Now I am at ease in my mind about Lidusya and Elza. It's a pity that Elza can only walk a little.

Sasha Lir's return is highly possible, since all of the Russian construction crews in Nizhni Tagil have been dismissed, and they send more Germans to us. Write to me, Mum. Your letters are very interesting for me.

Your son.

Monday 4th May 1942

Lidusenka!

I thought yesterday that there would be a workday on the 3rd of May 1942, and it turned out to be a day off, since another dispatcher came, for which

The railroad cars loaded with wood arrived at night. And again I had to run in order to warn the builders about the immediate unloading of these railroad cars. Then I went to find workers to load the bricks. At night the trucks arrived from different organizations to collect the bricks, but permission was not given to load them. Noise, crying, disputes, useless running of engines, etc. ensued.

Towards the morning I squatted in the clock house and fell into a doze for 20–30 minutes. Then I had hardly managed to plod to the barracks, when the assistant commander of the column presented the news: 'The commander of the column saw you sleeping at your post and he wants to raise a question about your discharge'. Apparently, this case will be forgotten.

Lidusya, I regret that you had to leave me so soon and that I didn't get the chance to talk to you much! But nevertheless I am happy about your visit. It cheered me up. Now I am almost confident that we will be able to meet in Nizhni Tagil again soon... If the present circumstances do not change. By the way, Yuri Veybert received a letter from his cousin-brother who was demobilized after the disbandment of a Russian construction crew in Krasnouralsk and now he lives at home.

Schumacher reported this interesting news. Today he met the business manager of the Sverdlovsk dramatic theatre, who arrived in Nizhni Tagil to chase them about the demobilization of Schumacher, having different papers from 'influential bodies'. The commissioner of our force replied that: 'All soldiers of our construction crew are subordinate only to Rappoport, the construction commander of the NKVD *Tagilstroy*, who is subordinate only to Beria and thence to Stalin'. Neither the military registration and enlistment office nor the NKVD can solve any problems of leave and demobilization of soldiers from our construction crew. Tomorrow this theatre director wants to obtain a personal appointment with Rappoport. I greatly want this matter to be resolved successfully for Schumacher. Indeed, he is 51 years old! Furthermore, of late he has permanently been ill. He has been vomiting constantly, a disorder of the stomach… If he stays here then poor Schumacher won't survive for very long.

Lidusya, I have still not undertaken anything concerned with the polarograph. It is probably also the case that passage to plant 183 can only be made with the permission of Rappoport. Perhaps, I should have a talk with the commissioner about this.

I wish the war would end faster and with a favourable outcome for our country. Reports from the Information Bureau about the offensive of our troops in the Kharkov direction are very consoling. I think that the superiority of the German force in the Kerch direction also will be liquidated by the pooling of our forces. Decisive battles begin. The future will show how this will be reflected in our life…

Our diet has changed slightly for the worse. For example, the porridge and that piece of fish that we ate together when you were here was probably the last for a long time. Now any cereals, sugar and fish are compensated for by bread. They give us 850 g of bread (instead of 700 g), a patty made with pea in the daytime and gruel with dumplings perhaps 2–3 times a day; sometimes they give us a pea skilly. Apparently, the pea porridge, fish, oat porridge and sweet tea with sugar have all disappeared from our menu for a long time.

During one of my night duties I was introduced to the new quality controller Rauschenbach;[30] his civilian speciality is as a scientific worker — an aircraft construction engineer. Now he studies higher mathematics with the Courant textbook. He is most interested in chemistry and physics. He is mentally alert, but his interests are spread too thinly in the different fields of science. This is the first person I have met who works to improve his qualifications. True, I forgot about the

---

[30]B.V. Rauschenbach was interred in the third wave of German deportations. He was a famous physicist working on the theory and design of rocket engines, ballistic missiles and was ultimately involved in Russia's space programme. He was interred after disagreements with the authorities, despite his prominence and importance to Russia's defence, under 'special conditions'. Armin later recalls that Rauschenbach did not have to participate in any hard labour associated with the brick plant, obtained a monthly wage from the Royal Institute and was permitted to submit one article per month concerning the theory of rocket engines and engines for space flight vehicles. Despite not having access to any reference works or monographs relating to this specialist subject, Rauschenbach was able to draw on the knowledge stored in his head and successfully submitted ten articles (written on his bed as he had no access to special writing areas, desks or equipment at the plant) to the Institute in the space of one year, thus fulfilling his obligation to them.

physicist Kromer who now studies with great enthusiasm the Pavlov and Semenchenko textbook on inorganic chemistry. I still work with the *Organic Chemistry* of Pavlov.

How did you get to Sverdlovsk, Lidusya? How did dearest Elza endure your absence? What are the prospects of a second visit to me? Let this visit not be taken into account, since we have seen each other for only a small amount of time. Would you have a chance to escape at the end of May? I await your letter soon.

Your Armin.

I received two letters from Nikolai Leont'yev in the reserve army. He is one of the middle-ranking officers in the zenith artillery battalion. He is safe and sound.[31]

Monday 18th May 1942

Lidusya, my dear! Before I send this letter, Dusya Isaeva delivered your parcel containing roe and fat — exactly one day before I would have run out of 'non-bread' additives, and I would have passed entirely over to the floury food of the builders. Many thanks for that. Also, please pass on my enormous appreciation to father. Fat is the very thing that is absent from our diet. You write that Kostya[32] brought the fat. Does that mean he is in Sverdlovsk? Does he bring any good news?

Dearest Lidusya, I have some news. They finished building the fence, the mushroom-shaped shed and the passage; they hung gates onto the hinges, and a big military guy was placed in the sentry box who lets no one get out from within the fence without being in formation. And another guy with a gun appeared under the mushroom. All this is done so that the special migrants surrounding us would not have an adverse effect on us,

[31]I was friends with Nikolai. He was among 25 chemists accepted into the UII in 1927. But from the third year onwards AGS chose electrochemistry, and Nikolai Leont'yev decided to be a paper producer, since his uncle was the director of a paper mill in the Urals. After the first course there were two months of military training. The command staff was trained for the artillery — AGS, 2003.

[32]Kostya — he was the elder brother of my stepfather A.V. Vorobyev, they lived in Ryazan — AGS.

as our commander Zakirov elucidated to us in an extensive speech during yesterday evening's roll call.

Your husband.

Lidusya! I know that it is too difficult to obtain fat in Sverdlovsk; therefore do not send any more!

Letter 17

Sunday 24th May 1942, 11.30 p.m.

Lidusya, my dear! Yesterday I received your postcard (dated the 19th of May) and the postcard from Mum. So, the 'planting' campaign is afoot. It's a pity that I cannot participate in it. I believe that it's difficult for you to be alone, and I suffer from the fact that I can't help you. How did you manage to obtain potatoes to plant? By the way, how much does a bucket of potatoes cost in Sverdlovsk? (I heard a price of 60 roubles. Is this correct?). Mum writes that dearest Elza has lost weight, but she is still healthy. Poor girlie! Now she goes without seeing you for entire days. Surely she makes 'pronunciations' to you for your long absences in connection with the gardening matters.

After the gardening campaign it is necessary to start another campaign concerning the stockpiling of firewood. What are the plans and possibilities concerning that? Mum writes that possibly you'll have to log wood right in the forest. Is this definitely true?

By the way, Lidusya, on the subject of firewood, I am reminded about winter and the felt boots. Many soldiers have already begun to worry about this and their home families have promised to send felt boots to them. If it is possible, could you send me some felt boots (probably for me to exchange?), since, apparently, it is still inevitable that I will remain in the construction crew for the winter (here or in another place).

Regarding the question of my remaining in the construction crew, hope has changed to disappointment. We were informed today that, after finishing the building work in Chelyabinsk, they dismissed a similar construction crew to our own. We have to finish building by the middle of June. However, it is more probable that they will move us on to another project. It seems to me that the latter outcome is more probable. In any case, our construction crew recently received a new complement of Germans who remained after the disbandment of the Russian construction crews. They even sent those Germans who had previously been allowed to work as civilians. They [the German civilian workers — GGW] had ration cards in Nizhni Tagil and they lived in private apartments after the disbandment of the Russian construction crew.

Monday 25th May 1942, 5.00 p.m.

This morning Schumacher's wife arrived, and then she left the same evening. Schumacher wanted to obtain a leave-warrant in order to walk with her through the city and to see her off on the train, but Nosikov (the squad commander) refused his request.

The presence of the fence and the clock house near our barracks has not limited the freedom of my movement within the territory of the construction crew, since I have a personal 24 hour pass. By the way, they fenced the 'mushroom', so that the sentry stands not under the mushroom, but sits in the fenced-in lodge.

Today is the first warm sunny day. I was on duty last night. The night passed comparatively quietly. The brigade of Remshevich has been sorting bricks wheeled out on trolleys from the furnaces. A locomotive arrived at night; it drove away with two gondola wagons of coal, and left one empty wagon to be loaded with lime. I ordered the 'gold brigade' to do this job. This nickname was given informally to the section of weak, young and elderly people who were assigned to us from the fourth column for auxiliary activities. They willingly transferred to us because the work is comparatively easy and we do not overburden them with regard to the setting of standard rates. But they work with some difficulty, since they are very exhausted. In particular, there are some difficulties with them at night. If you take your eye off them, even for half an hour, they crawl away like cockroaches to the warm furnace areas where they unload from, or they even sleep right where they work. Then I appear in the role of the terrible foreman and begin to 'administer justice and mete out punishment'. They creep out with moans from different nooks and crannies and barely begin to work again. These sections consist mainly of the Crimean and North-Caucasian (Krasnodar region) Germans. Initially, the NKVD evacuated these Germans over a period of a month with their families into Kazakhstan; then after 2–3 months they were mobilized (in January) and were conveyed for half a month to Nizhni Tagil, so that they arrived in the construction crew already very exhausted.

But on the other hand there is 'the brigade of Estler', which is the hope and the stronghold of our loading shop. They are healthy men, the Turinsk workers who still have some reserve of fat and food, and they load bricks and unload materials within a record time. For example, the brigade of Estler (6 people) can unload a gondola car of fine coke in 2–3 hours, and the 'gold brigade' (20–30 people) achieves the same task in 6–8 hours.

I have mastered my new speciality sufficiently well, but it is always necessary to be on the alert in order not to get sent to the guardroom. Recently they wanted to put away two people, but then let them go. One of them — a superintendent — left his work to take a break but left no one on the construction site; this was discovered by the authorities. But some extenuating circumstances excusing his fault were found and they let him go.

That's how we live, Lidusenka. Today I took a rest during the daytime. I studied organic chemistry, took a Russian bath, then played *gorodki*[33] and sunbathed. As you can see, it looks like a health resort rather than a construction crew. But certainly all this is laughter through tears. I would rather cheerfully take a part in the planting campaign or the raising of Elza (with discussions, drawing, music); I would rather sit with you comfortably on a settee, discussing the problem of firewood... All this can be, if the war ends soon and successfully. And it will not happen if the war is dragged out. But it can. Kisses to you and daughter.

Your HUSBAND, missing without a family.

Letter 18

Wednesday 3rd June 1942, 12.00 p.m.

Lidusya!

It's a long time since I received a letter from you. But I do not want to offend you, since I consider that you're in a period of harvesting madness

---

[33]A game similar to skittles.

Letter 20

Friday 19th June 1942

My dear!

Today I received a postcard from you (dated the 11th of June 1942) with the continuation where you describe two days of your life. It's hard for you to work in the university, take care of daughter, and keep a house — this burden is now placed on you. I'm very glad that your planting campaign is successful, and you have already dug up potatoes and planted the cabbage seedlings. The other day I received a letter from you (dated the 6th of June); you describe the difficulties connected with obtaining a pass for the UBAS dining room. I'm indignant at the behaviour of Dubrovskaya. I never expected it of her.

Some changes in my work have occurred in recent days. Now I am not a dispatcher of the loading shop any more, but a quality control inspector of the technical control division in the kilning shop. I do not know if it is a change for the better or for the worse. This change in my job is connected with a number of general changes in the workings of the brick plant. For failing to fulfil the 'May Plan' the director and the chief engineer were replaced. Now a soldier in our detachment, a building engineer, Popp, has been assigned as the chief engineer.

He has radically changed the way in which the unqualified manpower is used, and all former engineering and technical personnel were appointed to posts of shift masters etc. I'm among six people enrolled in OTK (quality control department). Today (the 19th of June) is already my third day in the new post. The organizational confusion has not been eliminated thus far. The completion of staff occurs; the laboratory is moved from one place to another (from a hut into the new room of the building office). Today I dealt with the loading and unloading of cabinets, tables, boxes, etc. from a cart for half a shift. On the whole I am contented thus far. I have read several books on brick production already. The process of brick production must be thoroughly investigated. My new job has not affected my nourishment thus far. As usual, I receive 700 g of bread and a second cauldron of skilly. I exchanged 100 g of bread for a bundle of green onions (for vitamin C, equivalent to 10 roubles). I took my collective farmer's boots (for the autumn and the winter) to be repaired: the new inner sole and the totally new sole will be made from a drive belt. I have to pay 500g of bread for the sole and the same amount for the work. I already saved a kilogram of bread, substituting bread by rusks. Soon I will have the new boots.

There is a lot of talk about home-leave for some categories of soldiers. First, the old men, older than 50 (born after 1891); second, boys younger than 18 years old (1925 and subsequent years). The medical board has been inspecting Schumacher for two days, and he expects to go home in two–three days. By the way, his wife and son have already left Sverdlovsk and gone to Tokmak, where the Odessa Theatre is assembled.

Remshevich hopes that they will let him go as he received an injury in the Patriotic War. Recently he was called into the staff office, and they questioned him in detail about his participation in the war. Boris Vladimirovich Getling (an electrical engineer from our seventeenth department of the third column) also expects that they will let him go home soon. He and his parents are orthodox, they are Russian according to their passports. He wrote many queries, and his former organization chivvied about his return. Recently he was summoned to an interrogation by the NKVD. They questioned him and said that his case will probably be resolved successfully and they will let him go home. He thinks that his former organization made a request to the NKVD in Moscow. I will clear up the case in more detail. It seems to me that the UBAS is barking up the wrong tree. Estler (a mechanic in a watch factory — he arrived in 1933 from Germany) was recently called into the staff office. His plant pleaded for his return, they are going to let him go back in two–three weeks.

Helmut Ioganovich Ryukkert received a notice that the Comintern and the International Red Aid are trying to release him. They enquired if he was still in our construction crew. Another note: after a medical examination they moved him temporarily into the weak command (he has lost a lot of weight and exhausted himself); also they gave him leave for ten days and an intensive diet (the same skilly, but the volume is one and a half times higher than a normal portion).

Thus far I work for 12 hours in one shift (from 7.00 a.m. to 7.00 p.m.) with a 12 hour break. This is much more tiresome than the break of 24 hours.

Usually at the end of the shift I do not go back to the barracks, but instead I make myself comfortable with a book on a pile of boards in the plant yard, and study till 9.00–10.00 p.m. It is irrational to come into the barracks earlier, since usually there is the evening trooping at 8.00 p.m., then a formation, drills, etc.

The construction of an artificial drying apparatus should be finished by the 1st of July, and then they can move us to another place. Today the chief of *Tagilstroy*, Rappoport, visited us. One of the soldier-builders said that Rappoport spoke to the soldiers and said: 'You will finish the drying apparatus, and then you'll be charged to start the building of a cement works'. Well that is exactly within my new speciality.

Lidusya, please write to me some more about our daughter. What questions does she ask? What phrases does she speak? What tricks can she do? Mum wrote to me that one time you searched for our daughter everywhere, but she had hidden under the bed. Here is a mischievous girl!

How did you solve the problems of getting firewood and the large room?[37] Mum wrote that the UII offered up some pieces of woodland 7 km beyond the Izoplit plant. Really, is it necessary for you to chop firewood on top of everything else?

Was a question about the passage of the Kroneberg family into the large room resolved? How do you contrive to live within your wage of 300 roubles?

I'll end here. Regards to Mum, to Dad, to Marina, to Granny Olga, to Nura, to Viktor. Kisses to my daughter and my selfless wife. Certainly, I would kiss you without the means of letter with great pleasure. You write that your visit here is now impossible. I hope that the circumstances will change.

Your husband.

Dear daughter!

How you do spend time in the kindergarten? You grandmother Madya wrote that you ran away home from the kindergarten. Do not do it, my

[37]The family were renting one large room in an apartment at this time which posed some difficulty in heating it throughout the Siberian winter — GGW.

girl, because dad is afraid that you might fall under a car. Make a picture for dad; show how you go for a walk along the street together with other children, hand-in-hand in couples. Please draw also how you and Mum do pick cabbages and carrots in the vegetable garden? Mum-Lydia writes that you dress yourself, you feed yourself, and you go to sleep yourself. Father is very contented that you are so independent and that you help Mum. Remember that it is now very difficult for Mum to live alone without Dad. Please try to help Mum in everything — to pick the vegetable garden, to keep your toys in their place. Your dad strongly kisses you.

Your dad.

Letter 21

Monday 22nd June 1942

My dear wife!

Since the two postcards that you sent (dated the 6th of June) I have not received any letters from you. I cannot be angry with you since I know that you twist in the whirl of life, and there is no time to think about letters, there is no time to write them. The question of firewood worries me greatly. What steps are you and Mum taking in order to store up winter firewood during the summer months? How can you get firewood? It would be preferable if the Kroneberg family could occupy the large room together with their own supply of firewood. Will our daughter go to a kindergarten dacha this summer? I hope that she is going to one, even so as not to have to spend the entire summer in the dusty city, so that she would gain strength for the forthcoming winter.

From your letter (dated the 1st of June) it is possible to conclude that you will have no leave over the summer. Please write in detail about your summer term: seminars, laboratory, examinations, yearly projects, preliminary examination and institute preparation courses, research, etc. If Elza leaves to spend some time at a dacha, then you will be almost entirely relieved of your domestic duties.

Do not worry about me. At the moment they have transferred me over to the so-called 'engineer's stove'. This means that I receive 800 g of bread, and I dine in a room for engineers, on an improved menu. This improvement is very essential both in the qualitative and in the quantitative sense. This morning there was a pea soup with explicit signs of meat present. For lunch we had soup and dumplings, again with some traces of meat; and a noodle soup without any meat was given to us in the evening, but with a bit of fat at the surface and plus one herring on top of that. I am excited about this menu after having been terribly bored with the meal skilly on which they have fed us since the 1st of May. One additional advantage of eating at the engineer's stove is the presence of individual coupons and civilized surroundings — tables with tablecloths in a separate room. Now there is no need to spend 30–40 minutes in front of the doors of the dining room amongst a crowd of those waiting for dinner, when it becomes the turn of the seventeenth department of third column to eat. Even then, after the department has taken its turn to sit at a table, we would have to wait a further 15–20 minutes for the waitress to bring the meal skilly.

If the standard of food does not deteriorate, then I will live magnificently without any additional 'domestic victuals'. Well, that is enough about food. It is very unpleasant for me that this matter concerning the engineers' nourishment took so much space in this letter. Obviously, I'm under the influence of the herring tail I ate recently.

As I wrote to you already, I now work in the post of quality control inspector in the kilning shop. Today is my sixth day in this job. Thus far I have to work daily from 7.00 a.m. to 7.00 p.m. Admittedly, this is much more difficult than the dispatcher's job for which I worked for 12 hours and then had a 24 hour break. The entire plant is only just beginning to adjust to the new regime. The laboratory was moved to a new accommodation block just two days ago. A question concerning how to sort the bricks was finally decided only today. From the 24th of June I will be working on the night shift for six days. I think that nevertheless this global organizational confusion will be sorted out soon and then the job will become much easier.

Together with the perturbations at work, the 'migration of people' is also occurring within the construction crews. Columns and departments are disbanded and re-formed anew according to the production needs.

This movement of people has not affected me yet, but they have already removed three men from our room.

[...]

This ill-fated war began this very day twelve months ago. Because of it we have been separated for three months already, but there is no end to the war in sight thus far... Lidusya, let's get ready to spend the winter apart. It will be hard for you in the winter after a summer without rest. Do not overload yourself with work, Lidusya. If you need money then it's better to sell or exchange some of the domestic jumble.

I hear nothing about obtaining any leave for anyone in the construction crew so far.

Well, goodbye, Lidusya. Anyway I hope that we will succeed in meeting together at least once more in the summer or in the autumn. Loving you — your husband — engineering worker!

Letter 21A[38]

Friday 3rd July 1942

Dear Lidusenka!

I write to you near the porch of the central post office of Nizhni Tagil whilst waiting for a one-and-a-half tonne truck,[39] whose brakes seized up two blocks before reaching the post office. I am riding in this truck in order to deliver 20 pieces of brick to the central laboratory for testing, and then we will go to the refractory plant to get 2 thermocouples (1.5 m in length, each set in an iron case). Three days earlier I drove (more accurately, walked on foot for 5–6 km in each direction) to the refractory plant and calibrated the thermocouples myself. I'm tormented greatly these days. For example, yesterday morning I went out wearing a coat and boots, and then it started to rain so that I got wet through. But I was not lost: I climbed up

[38]A postcard addressed to Sverdlovsk city, Chapaev Street, 8, flat 1. To Stromberg, Lydia Mikhaylovna.

[39]A 'GAZAA', a Soviet copy of the Ford AA — ASK.

to the Hoffman kiln and dried myself there for 15 minutes. The day before yesterday (the 1st of July), after going to the refractory plant, I visited Lisa Chernobrovkina and received a parcel from you and Mum. Many thanks to you both. In the past three days I have received a lot of letters: from Elga, Volosik, Adrian, Tosha, N.F. Lodygin, two postcards from you.[40]

Well I must finish this now. They have decided that, in the end, the truck has broken down and so we must go, probably, on foot to fetch the thermocouples.

Your husband.

Kisses to Elza. I await the arrival of Mum.

Letter 22

Monday 20th July 1942

Lidusya, my dear!

Already half a month has passed without receiving any letter from you. I am very distressed by the fact that I have had no information from you for so long. Did our daughter leave to go to the dacha? Has your crazy workload finished? How was the problem of firewood solved? Did Tosha decide that she is actually going to go to visit Oleg?[41] That is risky. Indeed, Oleg could be moved at any moment to some other place, and she will be left alone with their children amongst strangers. What news do your relatives from Krasnoufimsk send? How are they getting on there? How do they earn money? Probably, Lusia is out harvesting and Appolinariya Fedorovna cares for Vladimir's children. Where are Vladimir and Peter?[42]

---

[40]Elga — my half-sister; Volosik — Vladimir Krichagin, my cousin-brother; Tosha — Antonina Mikhaylovn, my sister-in-law; N.F. Lodygin — a colleague at the UCRI and old friend in our dancing circles — he and his wife were the most spectacular tango dancers whom I tried to copy very hard. To my regret, he became ill with tuberculosis and died soon after the war ended — AGS, 2003.

[41]Her husband who was serving in the Far East — EAZ.

[42]Lusia, Vladimir, Peter — Lydia's sister and brothers; Appolinariya Fedorovna — Lydia's mother — EAZ.

I worry about your finances. How you do manage to live on just 400 roubles? They don't actually pay any money to us, but in certain cases they can transfer money to families upon request, particularly if they need financial aid. I'm going to make this claim with an appropriate justification.

Currently my job is comparatively interesting, but very tiresome. I am a quality control inspector (since the 15th of July) in the kilning shop and am responsible for inspecting the finished products. I get up at 5.30 a.m., have breakfast at 6.00 a.m., and then I 'control' without a respite. I write reports, compose actions, check the furnace schedules, organize measurements of temperature and thrust, etc. Perhaps this is the most interesting work that I could undertake in the brick plant. It would be possible to make this work even more interesting, but I have a distaste for 'bricks'. I found an extensive literature on brick production in the refractory plant, but I read almost nothing from it. I prefer to read *Articles on the Physics of the Atom*. I get very tired. My feet are particularly sore, since I have to spend the whole day running around the furnaces and the sorting areas.

Tuesday 21st July 1942

I had to interrupt this letter because of the evening roll call. Our 'severe' commander of the column tortured us for an entire hour, shouting at four disruptive men. This morning I received a letter from you (dated the 9th of July), and a postcard from Mum later in the day (dated the 13th of July). I learned from these letters that our daughter has been on the dacha since the 5th of July, so that your workload has finally reduced; that in August the lumber works will start; and that you await the arrival of Lucy and the Valentina family from Tambov.[43]

I read the description of the nature and behaviour of Elza in your letter, and it seemed to me that you were writing about some other, entirely adult girl. I think that, even now, during these last four months (four months have passed already — how the time flies!) I find it difficult to know the mind of my own daughter: she is entirely independent, adult; and she discusses different themes.

[43]The wife and family of Lydia's brother, Vladimir.

From 7.00 p.m. I left to go on shift, and I will work on the night shift for a week.

By the way, an amusing event occurred yesterday. Ten days ago a geodesist, Shpurre, arrived here with one of the Sverdlovsk 'parties'. He has lived in our room for ten days since; I meet him daily, and we have talked together on several occasions. I only found out today that I had studied with this Shpurre for the entire year 1925–1926 in the ninth class of the school named after Lenin in Sverdlovsk city. In 1936 he graduated from Sverdlovsk State University specializing in astronomy and geodesy. He asked about Modestus and other our school mates (tell Modestus about it). Shpurre is familiar with M.O. Kler,[44] since they worked together in the department of geodesy at the Mining Institute. He knows Tatyana Steinberg. It was a very amusing turn of events. When he first recognized me, he asked if I had studied at the Leninist school. This is how life brings people together; even after 16 years of living in one city we had never met until now.

Yesterday a postcard arrived from Mum (dated the 19th of July), but it was written so indiscriminately (moreover the ink runs on this paper), so that I made out only about 70% of her handwriting, and 30% is still has to be investigated. In any case I understood from this postcard that you and Mum have begun to harvest firewood on the plots given by UBAS and UII, and that the large room has been rented by the Golubovsky family[45] with their own heating. I think that this would be a very successful solution to the problem of firewood in the winter.

It's 3.00 a.m. now. Dawn is breaking already. Soon I will have to make a circuit around the 'diocese': to check how the sorting goes, which raw materials are loaded into the furnace, etc.

I finish thus far with this: will this letter really be the last that you will receive in Sverdlovsk before your visit here to me? But could they suddenly refuse to grant you leave in August? It would be very sad if your trip does not take place, just like Mum's last mission. I await a recent letter from you with a description of your trip to our daughter at the dacha (I do

[44]The father of Modestus — AGS, 2003.

[45]The evacuated residents of Leningrad, friends of Magda Robertovna. She was a well-known pianist and her friend was a violinist — AGS, 2003.

not even know which dacha Elza is staying at). Has she asked to go home or not? Did she cry when you left? Or maybe she likes staying on the dacha and she let you go without any tears? Is she healthy?

I would very much like to have a photograph of my daughter! If you do not have a new one, then please bring at least something from her earlier photos. When our little girl returns from the dacha please take a photo together with her, because I cannot even visualize what my daughter looks like after four months of being away.

See you soon, my dear. I hope that I will meet with you at any rate.

Your lonely husband.

Sunday 28th July 1942

Lidusenka!

One additional practical observation. I already wrote you that it is possible to obtain felt boots here. The prices are measured in tobacco, schnapps, bread and money in the following ratio: one container of low-grade tobacco = 0.25 litres of schnapps = a loaf (1.5 kg) of bread = 200 roubles. These three objects in essence substitute money and are the means of circulation. Prices are approximately such: completely shabby felt boots (without a sole) = two containers of low-grade tobacco; boots with overshoes (used) = 45 containers of tobacco = 800–1000 roubles; almost new felt boots = 45 containers of tobacco = 1l of schnapps; a bread ration of 800 g = 100–110 roubles; a shirt = 70 roubles; some long rubber boots = a bundle (100 g) of light tobacco = 350 roubles; etc.

I don't want to sell my boots with the overshoes. Therefore I beg you, Lidusya, to bring one or another or the third of the listed prices. Certainly it is far too difficult to obtain tobacco in Sverdlovsk. It can be purchased here for money. There is an excessive demand for schnapps. It would be perfect if you could mange to obtain four or five 0.25 litre bottles. If you have some rusks, please bring them with you. If it's impossible to obtain tobacco and schnapps then bring money — around 1000 roubles (sell some junk). This will be enough to solve the problem of felt boots…

Strongly kiss you, Lidusya, and I await your arrival impatiently.

I will hopefully receive another letter from you soon. Will it arrive or will it not? This thought haunts me throughout the entire day. I recently sent you a telegram (on the 5th of August) so that you may arrive after I get the felt boots. Forgive me for this small craftiness. I have already purchased the felt boots for you, and our shoemaker Brouman soled it perfectly. However, I will try to send the boots to you at the first opportunity. Nevertheless I decided to send the telegram. Since I completely understand that you, just as Mum, are busy going about your daily life and work, and that you are very tired and greatly weakened due to poor nourishment and overwork. And at this time I decided that the felt boots might possibly be that small additive, ΔP, which will tip the scales in favour of a trip to me, balanced thus far by the various pros and cons. Again please forgive me, Lidusenka. If you are not able to arrive, then of course I will send the felt boots to you with the first 'Sverdlovsk wife' (if she will agree to take them). But if you do not arrive during August, then I'm afraid that we won't meet for the foreseeable future. Since, on the 1st of September, the semester will begin at the UII, then the harvesting campaign (tending your own vegetable garden, trips to kolkhozes, working on how to store the vegetables). However, on the basis of current events, they may now send us somewhere further into Siberia to the north, most probably at the end of the summer season at the brick plant. Well, let's not guess at the future, we'll wait and see.

My life passes without any particular changes so far. I'm a quality control inspector. This week (the 3rd to the 9th of August, inclusive) I'm on the day shift; the following week on the night shift (the 9th to the 16th of August). I look forward to switching over to the night shift. Night work is much calmer; there is no need to be present at the evening roll call; there is a possibility to eat for a fourth time — the night skilly served for the night shift between 1.00 and 2.00 a.m.; in this case there is an unwritten tradition that the person on duty gives engineers the remaining leftovers without taking any coupons.

Evening roll call is especially onerous. I go to work at 6.30 a.m., and I return to the barracks at 8.00 p.m. You've just managed to sit down, when the hateful exclamation of the person on duty is heard: 'Get out on the evening roll call!'and this check always lasts for at least 1–1.5 hours, since our commander of the column (Zakirov, you know him) loves gabbling, and he speaks Russian very badly (he's a Bashkir) and he always speaks about the same things: poor discipline, insufficient cleanliness, bad mopping of the floor... And having been running around the plant for the whole day, you can hardly stay on your feet, and want only to sleep; and you think: 'Is this check going to finish soon, is this life going to end soon, it is ever going to be possible to begin a new life together with my dear wife, my daughter, with relatives?' The only free hours are during the daytime break. The past few days I have used this interlude for carrying out extensive repairs to my trousers. I have sewn big patches on the left-hand side, and darned the right-hand side. The trousers (black with a white stripe) came out well, almost like new! I have also learnt how to darn socks (on a wooden spoon) and underwear. When I return home, you will receive a first-class husband. You will be occupied with scientific works for your candidate thesis, and I will be washing linen and darning for the entire family. Will this be soon?

In essence I am mainly occupied with the sorting of bricks, by their inspection, grading, marking and sampling from the various batches, with controlling the loading of bricks according to types etc. The technological process of the kilning of bricks is out of my sight, because of a lack of time and ... enthusiasm. Finally, the malnourishment, lack of sleep and my oppressed state of spirit are beginning to show. 'I'm not the one I used to be'. –Grigoriy Gryazny's words from *The Tsar's Bride*[47] spring to mind when I compare my fitness for work between the periods at UBAS and here. There's no need to say anything more.

The trend in our nourishment has turned for the worse again. I still receive coupons to the engineers' cauldron as before, but the cauldron itself has become somewhat different. The content of meal products was reduced by 50% (noodles, dumplings, croup, peas) in order to replace these with vegetables. The traditional evening roll call was cancelled, and the skilly has become a very thin liquid. Any vegetables (cabbage, mushrooms) have not ripened yet, and therefore they are added, probably, in scant quantities. We hope that the trend in our nourishment will go upwards again with the appearance of fresh potato on the market and by a general improvement in the situation. We still receive 800 g of bread a day.

---

[47]An opera in four acts by Nikolai Rimsky-Korsakov.

I exchanged a piece of bread (100 g) for a bundle of green onions and 'silo' (a bunch of vegetable tops of beet, cabbage, dill, pairs of onions and carrots) from women who come with baskets to the dining room. Unfortunately, the commander of our detachment often turns these women away, and therefore it is not always possible to obtain onion and silo.

Sasha Lir has returned from Shaytanka (a sand pit) to the brick plant and he now works unloading the bricks from the furnaces (an unloader, hard work). I want to attempt to arrange for him to be transferred to us in quality control to the post of temperature controller.

I sorted out the problem concerning my timesheets only recently. I already wrote to you that they have recently overcharged me for April, May, June, only posting the deductions (food, loan, lottery, penalty fines!), but they did not add on the payroll, since the chief of the loading shop did not fill in the timesheets for me (and other dispatchers). The problem was complicated by the absence of the chief, who has left to oversee the hay mowing with a group of soldiers near Ufa. They did not want to sign the timesheets; there were many troubles, scurrying around... But now everything is sorted out and they promise to put me back on the payroll. As a dispatcher of the loading shop I had a salary of 350 roubles, and now it's 400 roubles. It's a good idea to ask how much money I have on the account currently. By the way, speaking about the penalty fine, at the end of July an order appeared, that reprimanded me and some other dispatchers, and deducted a material fine from the plant because some railroad cars were left idle for a period of time; it exacted 224 roubles from me and 220, 500 and 800 roubles from some other dispatchers. It's a deplorable fact. But accepting the fact is probably the only option we have. I spoke to the director today. He says: 'It's useless to write anything without finding new guilty parties, whom I could charge instead'. The railroad cars were indeed left idle, since the previous director moved a good brigade of loaders to the furnace, and then only gave a weak brigade to the loading shop ('weaklys' or 'sicklys' on the local jargon, they can hardly drag one foot in front of the other). It is clear that 30 people from the 'weakly' brigade could not unload a railroad car containing fine coke within 12 hours, which was unloaded previously in 2 hours by 8 healthy loaders (they receive parcels from home frequently). However, the size of the penalty was determined by the fact that once, in June, I signed in a bundle of invoices for the railroad cars, which were unloaded at the plant over that period of days (not only on my shift, for sure). That's how I came to be punished for my inexperience. But this is merely a trifle, since it is unknown if I will ever obtain any wages in any case. And that's nothing to speak about if we compare it to the background of these general events and everything that could possibly happen to me. It is a shame, only because of my inexperience, and it is unpleasant that they issued a reprimand…

Mum writes that you collected six cubic metres of firewood. Now all that remains is how to transport it. How is Mum getting on with the chopping of the firewood? Why are only Mum and Marina going to do it? What about Dad and Nyura? Have you seen Helen Shubnikova? What does she say about Ashkhabad? You write that you chopped wood together with Volkov and Telejnikova. Send them my regards for me. Did

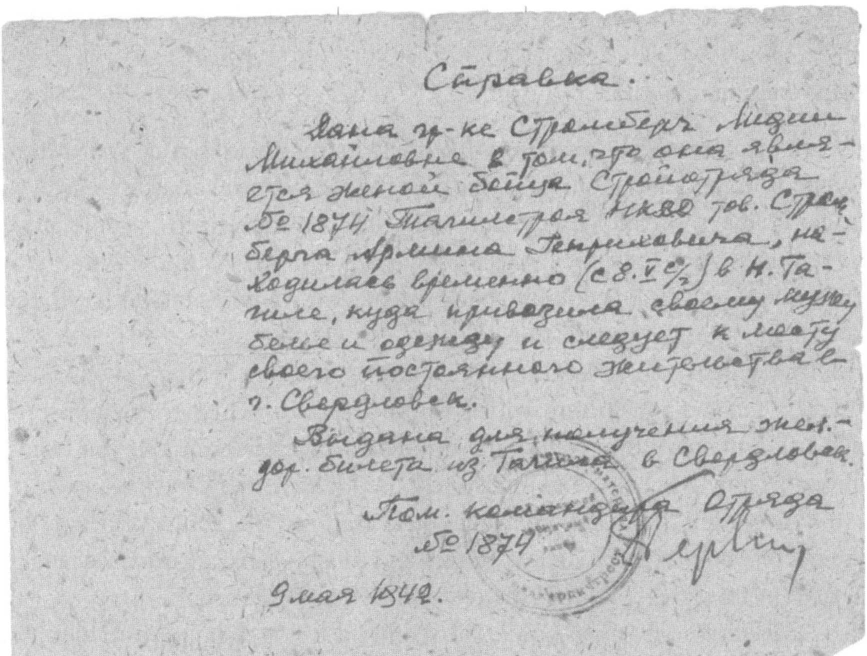

Справка.

Дана гр-ке Стромберг Лидии Михайловне в том, что она является женой бойца Стройотряда № 1874 Тагилстроя НКВД тов. Стромберга Армина Генриховича, находилась временно (с 8.V с/г) в Н. Тагиле, куда привозила своему мужу белье и одежду и следует к месту своего постоянного жительства в г. Свердловск.

Выдана для получения жел.-дор. билета из Тагила в Свердловск.

Пом. командира отряда № 1874 [signature]

9 мая 1942.

One of two surviving certificates granting Lydia permission to buy a train ticket to visit her mobilised husband in N. Tagil.

you see our daughter on the 2nd of August? I await your telephone call on the 10th or 11th of August. Sasha Lir recently came to call, he brought a letter from you (I finished reading this letter earlier in the evening in the barracks). I'm happy for you that the vegetables are ripening nicely. I hope that the children will not steal them.

You write that your leave was revoked, and the planned mission to visit me is cancelled. Does this mean that there is no hope whatsoever of your arrival?

Well, I'll end here, since there will probably be an evening roll call now.

Love to you and our daughter.

Your Armin.

Authors' note: L.M. visited Nizhni Tagil between the 16th and the 21st of August. She carried this next letter to Elza.

Friday 21st August 1942

Dear daughter — Elza!

I send this letter to you with your Mum, and strongly kiss you. Your Mum, Lydia, told me a lot about you. She told me that you have become grown-up and clever. I would like to see you, but I am mobilized into the army now, and they do not let those mobilized go home. As soon as the war ends, the day after, I will come back to you and we will go together to the zoo or the recreation park to gather bits for the samovar.

I have sent you several toys: two sieves, through which you can sow sand or earth; a mushroom with a removable cup, which you can use to put different small things into; the figures of different animals. You surely already know what these animals are called. Draw them for me.

Write a letter to Dad, dear daughter. Write to me about your stay on the dacha. How did you gather berries and mushrooms? How did you take a walk in the forest? Did all the children listen to the teacher?

Dear daughter, you are older now, and I ask you to help Mum with the housekeeping. Help Mum to wash potato. Do not throw your toys about. When you finish playing with your toys, then put them all back into the box.

A sketch drawn by Armin depicting his daughter picking mushrooms.

Goodbye, dear daughter. Soon Dad will write another letter to you. And please draw some different pictures for me.

Your dad.

Letter 26

Sunday 6th September 1942

My dear, sweet Lidusya! My dear wife!

I wanted to hear your voice yesterday on the phone so much — and the conversation failed! It miscarried because of a foolish chance. Once in a letter you gave me an incorrect phone number for Ilin: 5608. But then when I shared my sorrows with Sasha Lir, he gave me the correct number — 4507. You increased every single digit by one, and I ended up calling an office of a metal work plant; certainly, the conversation with this random person gave me no pleasure. I did not write any letters to you in the hope of having this conversation… Fortunately, on the 8th of September our head of department, Rita Mikhaylovna Vulakh, will go to Sverdlovsk.

She was in the office when I uttered howls of frustration at this ill-fated call and slammed down the receiver. She promised to ring you on the telephone Shlygin (4507!) so that you would visit her to receive this letter. She will tell you about our job. Furthermore, she promised to bring some salt for me from you; if you can, please send me 300–400 g of salt. Please send some writing paper or notebooks as well, since my last reserves of writing paper are running out.

I prepared a long list of questions and news for yesterday's conversation. Now I have to present everything in the letter, and to wait for the answers for another 7–8 days (Vulakh should return on the 13th of September).

Thus, Lidusenka, first some questions.

1. Do you have any health problems? How are Elza and relatives — Mum, Marina, Grandfather Sasha, Grandma Olga, Viktor?
2. Have the living conditions changed after the Golubovski family's arrival? You wrote that Marina sleeps in your room now. How are the others accommodated?
3. Is there any possibility of having some firewood delivered? This issue worries me. Really, why do they not agree to transport it even in exchange for the felt boots?
4. The garden: Have you begun harvesting potatoes yet? How much was stolen? What is the market price? They are already digging potatoes around here. The price is 130–150 roubles per bucket.
5. Your job: How are the examinations going? When does the next semester start? Will you be able to have a little rest during September?
6. Harvesting campaign: Did they assign you to this task or not: kolkhozes, vegetable storage?
7. Nourishment: Did UBAS give you any dining rights in the academic dining room for September? Did you purchase products and goods on the cards in August?
8. How about the footwear for the autumn and winter, for you and for our daughter: autumnal shoes and overshoes for you, felt boots — for our daughter?
9. Family: Is Modestus with Irina? Adrian with Evegeny? Tosha with Michael and Leo? Nina with Evgenya Samoylovna? Karpachev? Rempel? Zelyanskaya? Parysheva? Kuzma? Lepik?

Now I will describe my life. My job is going satisfactorily. The unfortunate report of our chief engineer fell into oblivion, since the kilning of pots turned out to be quite successful. There were no other incidents. I and Kromer decided to switch to the daily shifts: 24 hours of work and 24 hours of rest. Since, officially we have a ten-hour working day, thus we have four hours of official breaks. We use these to rest during less intensive times. This measure has increased our amount of free time. Today, as I write, is my day off. I go to work at 8.00 p.m. on the 6th and finish at 7.00 p.m. on the 7th. I studied and made an abstract of chapter three of *Articles on the Physics of the Atom* during the morning, which I had neglected during August; and after dinner I set about writing you this letter. Sasha Lir works alongside me during this shift; I arranged a job for him as a temperature controller: he measures temperatures, writes out the routines for the furnaces, marks bricks. Generally his job is easy. But Sasha, apparently, is a bit of a pessimist, and therefore he is to be permanently found in a gloomy state of despondency, claiming that the quality control is pointless and so forth. Of course, there is a reason for his gloomy mood. His only source of warm clothing is an overcoat — almost without any back to it — and tarpaulin shoes; he has neither a quilted jacket nor felt boots nor a warm cap. In spring he sold his padded trousers and padded jacket (and sent the money to his mother), and now he is left without any warm clothing. I must find a way to help him somehow. Please find Vera's address and write to her about this (she will stay in the kolkhoz in the Manchazhsk region for a further two months). Maybe she will be able to provide some assistance to him. For example, if he had a little money or an 'exchange fund' he could acquire warm clothes here.

After your arrival my nourishment has been OK. I twice bought potatoes and other vegetables. One time it was a half a bucket of potatoes + 30 onions + 10 carrots + 5 beets + 3 turnips, etc. I bartered this with one tender-hearted woman for 500 g of bread + 100 roubles. Another time I managed to purchase a bucket of potatoes + 20 carrots for 150 roubles, so I am eating a lot of vegetables: I make boiled potato, puree, preserved food, soup. Actually, sometimes it's burnt, sometimes I over-salt it, but generally I eat up all these dishes with great pleasure. The food in the dining room has also improved over the last four days. After two weeks of eating oat-porridge without any salt three times a day (sometimes

90–95% *aqua distillata*), they began to feed us some soups made with potato, cabbage, meat, fish and … salt.

Though, all of this was superfluous to me, because of your care, my dear. Your money matters disturb me greatly. Indeed, you owe 700 roubles to my Mum because of me. Soon I will be able to help you a little. Permission has already been given for the transfer of 75% of my wages earned over four months (April–July) to you, which is 500 roubles, and they paid me 170 roubles today. I earned 1470 roubles for the above-mentioned period, but 790 roubles were deducted, including:

— four months' income tax	35 roubles
— four months' culture tax	31 roubles
— four months' military tax	150 roubles
— a loan over four months (I signed for 400 roubles)	120 roubles
— the lottery (I signed up for 100 roubles)	50 roubles
— four months' tax for childless couples	74 roubles
— three months' food bills	240 roubles
— food in July	90 roubles
Total amount	790 roubles

By the way, please send me a certificate about 'our family size' from the housing management, since they consider everyone to be childless who does not have this information.

That's my money matters. I obtained the first set of very poor felt boots for some tobacco, and then other goods. The shoemaker will sole both pairs for 300 roubles (I have already given him 200 roubles). He will cover the one pair of felt boots that has only skin on the bottom with a rubber sole. The end product will be some sort of water-resistant felt boot. I will try to send them to you for the exchange fund. The second pair of felt boots is almost identical to those that I gave to you, so that I will have warm felt boots in the winter.

I already sold one block of 'butter' for 300 roubles, soon I want to sell the second. Potatoes become cheaper every day (today they offer a bucket of small potatoes for 130 roubles). After the potatoes have been harvested and stored in the cellars, they say that the prices will increase again.

Therefore I want to build up a stockpile of potato consisting of at least 5–8 buckets! I will attempt to barter this for a small quilt, for some bread, and for some money. Incidentally, the price of bread has fallen significantly, and they are now unwilling to exchange bread for potatoes (800 g of bread gets one only 3–4 kg of potato, and 1 kg of bread costs about 80 roubles.) But where could I store the potatoes? I want to make a box under the bunks and fill it up with potatoes. Here are the problems that your husband-scientist intends to solve!

They are gradually supplying us with clothes. Rubber boots are available for cash (for 45 roubles); they gave us some linen — drawers and a jacket made of a very good, thin and durable material. They only gave me a pillowcase as bed linen (they also gave blankets, towels, mattress covers, and sheets to others, depending on their needs). I am missing some things (for example a mattress). However, it's not a big disappointment because this bed cloth is only given for temporary use.

They have begun to give out outer clothing too: Estonian jackets and trousers. I do not know if I'll manage to obtain these.

I have several requests and instructions to you. 1) Please send the receipt about the transfer of money to the wife of Fridrikhsen, if you did not lose it yet (give it to Vulakh). 2) Get my literary catalogue on polarographic analysis from Al. Iv. Zelyanskaya. 3) See Veniamin Yakovlevich Vlasov in the UBAS and ask him if he received my letter and did he send it further to Tomsk with the addition of those materials, which I did not have here (he requested a description of figure three).

Well, that seems to be all the tasks. This letter turned out to be very business-like, since I wrote it on the basis of our cancelled phone conversation yesterday. One additional deal. Let's arrange the next phone conversation on an even day of the month. For example, I will order a telephone conversation 'with the pre-request' (4507) on Tuesday the 22nd of September at 9.30 p.m. If you're busy this evening, then send me your 'counter plan'. All my evenings are free, but I'll be on duty in the evening on the even days and this is more convenient for me to call you.

It is a pity that our conversation did not happen. Mainly I wanted to hear your voice, your stories about our dear and glorious daughter. Well, it doesn't matter. On the 13th I will probably receive a letter together with

Vulakh, in which you will tell me about your life. Nevertheless, we should arrange to talk on the 22nd of September (or another day).

I strongly, strongly kiss you my dear, and our dear Elza. Thank you for your last visit. It is extremely supportive for me, both morally and financially. Now I am ready to look the future in the face, although it promises to hold nothing of consolation thus far.

Your husband.

Letter 27

Thursday 17th September 1942, 4.00 p.m.

Dear Lidusya,

I received your letter yesterday, together with a parcel and numerous illustrations of Elza and Viktor, delivered by Rita Mikhaylovna. Many, many thanks to you. It brings me shame to see you send those pitiful crumbs to me, which were given out on the ration cards to you and our daughter. Do not do this any more — especially because I eat well until now. It is terrible to recall now how I lived with only bread and porridge before your arrival. Now I am in bliss. I eat potato in all forms: boiled, baked, fried and stewed. And it's not only potato. There is cabbage, beet, carrot, turnip and onion. A little of everything, and what a remarkable preserve I am making! It's a pity that I cannot entertain you with these outstanding dishes! I barely eat bread now. I grew tired of it this past time, and it is preferable now to exchange bread for potato, since, for a bread ration of 800 g one can obtain 3.5–4 kg of potatoes. I do not really know what my main speciality is: a quality control inspector or a housewife. The latter is more accurate, since I spend more time obtaining, cooking and eating food, than I spend in the job of the quality controller.

I will tell you about my economic successes. The first — obtaining vegetables. First, I bartered our daughter's small quilt for two large buckets of potato after a long walk around the settlement and the barracks (i.e. for 300 roubles, since a bucket of potatoes costs 150 roubles, on average). It seems to be quite a successful exchange, since I made the rounds of 50 houses. In the beginning my asking price was six buckets (!) — they ignored me. Gradually I decreased my demands to four, three, two and a half, and two buckets. In the end, they offered one and a half buckets in one of the barracks. I did not agree and left the room immediately. Then a lady overtook me and said that she had discussed it with her husband and decided to offer me two buckets of potatoes in exchange for the quilt. An amusing episode occurred at that moment. During the conversation I mentioned that my family lives in Sverdlovsk, and the lady began to worry: 'Why would they give away this quilt?' Certainly, I tried to set her at ease, and assured her that this quilt was not stolen. Nevertheless the lady looked sideways at me warily: 'Anything could happen when you have a deal with an unknown person'. Obviously, my appearance was not credible... Then I purchased one bucket of potatoes in the settlement from one little girl for 120 roubles. I admit that I swindled her a little, telling her that a bucket costs 100–110 roubles on the market (this trick didn't turn out well in other places). I bartered the fourth bucket in the settlement for 1.5 kg of bread (two rations), moreover in addition they gave me the head of a cabbage, six carrots and a bundle of green onions. On top of that I purchased another large head of cabbage for 20 roubles. On two previous occasions, I have bought another half-bucket of potatoes, with different additives in the form of carrots, beet, etc. At present I have three and a half buckets of the potato (I ate one and a half buckets this month) and have 430 roubles still available (I still owe the shoemaker 100 roubles for the repair of two pairs of felt boots). More experienced people say that soon potatoes will begin to increase in price. Therefore I am going to purchase another two buckets of potatoes with the remaining money. By the way, it is not easy to purchase potatoes even now. The majority of households refuse to sell them, reminding me that this is their only food product for the entire year. Still, some time later on it will become extremely difficult to purchase potatoes. I still have some bread, and I will exchange it for potatoes. That's how your husband has gained considerable economic skills!

But even greater abilities have manifested themselves in my preparation of all possible dishes made from potatoes. Some dishes could be considered somewhat exotic. For example, a potato baked on the brick-roasting Hoffman kiln No. 1, Hoffman No. 2 or the Buhrer kiln No. 3! Depending on the desire of the customer it is possible to bake potato without a browned crust, with a browned crust, or with a slightly charred crust. I will describe to you, for example, the prescription of the preparation of potato with a browned crust. Ten potatoes of average size are strung together using a 1.2 mm diameter copper wire 1.5 m in length. The potatoes are then placed into the small fuel tube in the fourth row behind the fire at a depth of 1m, and there they are cooked for 15–20 minutes, depending on the size of potato. On the second row behind the fire potatoes with a charred crust are cooked, and the usual crust can be made on the sixth row from the fire. Generally the matters concerning the proper preparation of baked potato in brick kilns are very poorly described in the literature until now, and it would be possible to write a brilliant PhD thesis on this theme. All the people in our room are enthusiastically occupied with these culinary experiments. For example, today our three lieutenants in the room — Baytinger (the manager of the joinery), Brotsman (a dispatcher in the loading shop) and Depershmidt (a driver) — prepared a dish *'kartoffel und kleiser'*, which is 'potato with dumplings' translated into Russian. Now (in the evening) the other two drivers prepare fried potato.[48] I cooked borscht made with cabbage, beets, turnips, carrot, potato and onion, seasoned with semolina. By the way, I revealed some strange lumps in my last package, and after more detailed study it proved to be, according to all their properties, similar to a product being produced back home in large quantities. Lidusya, thank you for this surprise.[49]

The third skill I have acquired is in the eating of things. This process takes a lot of time, since in the time between being served porridge dishes, I use every free hour to go home to cook different dishes, and I eat them, or I go to the Hoffman kiln No. 1 and make the aforementioned experiments with baked potatoes and my subsequent engorgement on these. I never thought that the issue of nourishment would play such a significant role of my life. I had always lived in 'the sphere of pure science', and these types of life skills that I have learned will be extremely useful to me in the future.

---

[48]Father later recalled a story about eating the leftovers in the lieutenants' frying pan, and the ensuing scandal this caused. It resulted in a meeting in the engineers' room organized by Rauschenbach concerning the avoidance of similar actions in the future — EAZ.
[49]Possibly, he is describing the inclusion of some sugar or chocolate — EAZ.

However, all my 'troubles' must be a trifle in comparison to those difficulties that you encounter. The matter of firewood is especially troubling to me. It's a pity that we do not know any friendly driver in Sverdlovsk. At present, two drivers live in the same room as me, and they often deliver firewood 'along the way' for friends and family. Maybe there is a driver who is familiar with some of our friends? To my mind, this is the only solution.

Furthermore, there is the question of potatoes. It's the third day that it has rained in Nizhni Tagil, with wind, slush and even snow today

A scan of a letter written by Elza to her father with the drawing of a house.

(the 17th of September). Surely you have the same weather in Sverdlovsk, and your potatoes are still in the ground (in the letter dated the 13th of September you write that only 30 kg was brought home). Now you have to wait for a break in the weather. I hope that the rest of the potato crop will not be stolen from the field in this time. You write that the potatoes in the vegetable garden are diseased. What does this mean? Worms? Did they rot from the damp? Are they good enough to be stored over the winter?

You do not have autumn footwear and overshoes, and you're too busy to do anything about it. My poor wife! The concerns about our daughter, the family, and the idler-husband do not leave you much time to think about yourself. Lidusenka, I strongly object to this. Please obtain footwear and boots or overshoes for yourself at any cost! I can't imagine how you can walk outdoors these days (the 15th to the 17th of September), if you have the same weather in Sverdlovsk as we have here. I am in bliss walking in the rubber knee-boots. I slap 'straight' through our terrible mud, and my feet are warm and dry inside.

I'm very worried about the 'behaviour' of our daughter. Her ears are infected, and sores have appeared. That's how our daughter 'got well' in the summer. Write to me in detail about our daughter's ear infection. What is this? Inflammation of the middle ear? And generally write in details about her. I 'examined' her drawings thoroughly, and I came to the conclusion that she had drawn a house (the window is drawn separately), and a bed with a blanket and pillow. I could not decipher the upper four figures. Please ask Elza to draw pictures especially for me and describe them to you.

Our life is comparatively quiet. News breaks that five old men (who are older than 50 years old) and three youngsters (younger than 18 years old) have leave to 'return home'. Poor Schumacher was not included in this list for some reason, although he is already 51 years old. I feel sorry for this poor old man, but maybe this misunderstanding will be sorted out. They are gradually sending 'specialists' to other enterprises. Turners, metal craftsmen, carpenters and drivers have been removed thus far. There is a demand only for the mechanics, the builders and the electricians among the engineering staff. There is no demand for the chemists and the physicists yet. They temporarily removed Rauschenbach yesterday for two weeks (together with another 100 men) to finish one building. They also wanted to take Fridrikhsen to be a supply worker for the sand quarry (without a doubt, he did not agree to this request), but Vulakh strongly objected and refused to let him go. That's how we live now, and it's hard to guess what tomorrow will bring in our uncertain life.

Sasha Lir works as a temperature controller together with me. He is gloomy, melancholic, hungry and without any warm clothing. I try to feed him up with potato sometimes. Yesterday he received a letter from Tyumen with a note from his mother, saying that she had sent a padded jacket, padded trousers, and his brother's boots to him via a friend. I'm happy for him, because he has suffered from the cold in recent days, being without an overcoat and only having tarpaulin shoes.

Lidusenka, please write to me more often, I am envious that Fridrikhsen receives a letter from his wife every 12 days. True, they don't have any children. I allow a great deal of leeway for the difficulties you have in caring for Elza, and I just ask you to write to me once every five days. I also promise to write frequently. By the way, Fridrikhsen is very disappointed. His wife has now taken in some relatives evacuated from Nal'chik to live with her: her mother and sister-in-law with a four-year-old son. This really aggravates her situation and limits her capabilities.

I will make a phone call to Modestus on Thursday the 24th of September at 9.30 p.m. If you are unable to come, then I will at least have a chat with Modestus.

I was struck by the news concerning the family squabble between Adrian and Zhenya. I always knew that he was a self-centred egotist, but I never thought that the material difficulties he is experiencing would make him behave in this way. I certainly think that they will not come to a divorce. I agree with you that the biggest source of these family discords (if not the main fault) is caused by mothers-in-law, fathers-in-law, etc. Fridrikhsen told me several histories of his 'family chronicle' when the situation came close to a divorce as a result of the mother-in-law stirring up her daughters against her sons-in-law (another husband and Fridrikhsen's brother). In his case the mother-in-law lived separately in other house, but in the same village.

Thus it is true, Lidusya, that we must live separately from all our relatives. Listening to all these histories, I have come to the conclusion that my mum appears to be a good mother-in-law, but… Nevertheless it will be better if I live together with you and Elza, separately, if only Mum's material needs would enable her to live separately… It's very difficult for her now because of the arrival of Granny Olga, Viktor and Nyura. Please help Mum as much as you can. Kiss you hard.

Your Armin.

Letter 28

Sunday 20th September 1942

My dear sweetheart!

Tomorrow our soldier Eugene Mikhaylovich Schumacher is to be demobilized (!) from our construction crew *forever* (!!), and he leaves to Sverdlovsk at first, and then on to Tokmak to his family. I envy him sincerely. He was demobilized because of his old age. I'm happy for the old man. He is amazingly cheerful for his years, and even his half a year internment in our camp did not wear down his vitality very much; of course, he has weakened during this time. In recent days, Schumacher was working in complete distress, since they demobilized five old men (who were more than 50 years old), and they missed him out. Happily, this misunderstanding was soon corrected, and he could have left even as early as this Saturday, had not one of the workforce fallen ill on that day, who had to sign some of his release papers. Thus, his departure has been delayed until Monday the 21st of September (and maybe even longer).

I gave him some money for the train ticket, and he promised to carry a letter to you in return for this, and to pay back the money to you. I should mention that recently they have stopped giving money to the soldiers of our construction crew, which was previously received via mail or telegraph, and they have instead put this money aside 'into a current account', so please, do not *under any circumstances* make any money transfers to me (and there is no need now).

My life has been very quiet during these past three days, without any changes. On the 18th I bartered with one old woman who favoured me — babushka Manya — who gave me a heap of vegetables for 1.5 kg of bread (two rations) plus 100 roubles in cash. Namely, I received one and a half buckets of potatoes + two cabbages + three large turnips + six beets + a kilo of onions + a bundle of green onions + dill. I consider this to be successful '*gescheft*', since the market prices in Nizhni Tagil are astronomical: 1 kg of turnip-shaped onion — 70 roubles; the head of a cabbage is 30–40 roubles; 10 carrots — 13–15 roubles; 3 beets — 10 roubles; a bucket of potatoes is 150 roubles. But 1 kg of bread costs about 100 roubles. I will save three additional rations today (2.4 kg) and tomorrow I will go again to change them for potatoes. In the last three days I only ate 800 g of bread, but then I eat an enormous quantity of potato and other vegetables, and therefore I'm completely satisfied. It is profitable to exchange potatoes for bread under our present conditions in every sense, including the scientific considerations expressed by my neighbour Paul Emilevich Rikert. One kg of bread has a calorific value equal to 2.5 kg of potatoes, and in exchange it provides 4–5 kg of potatoes. At the same time I try to obtain some additional vegetables during the exchange and this is much more advantageous than to buy vegetables separately. Today I made a big business deal — I put together a large box for the potatoes in our joinery office (holding 6–8 buckets of potatoes), brought it home, installed it under the cot and filled it up with vegetables. It turned out I have approximately two-thirds of a box of potato (i.e. approximately three and a half buckets) already. But when I put in all of the other vegetables, the box was filled right up to the top. I want to increase my reserves of potatoes up to 6–7 buckets, mainly by exchanging them for bread.

They still feed us badly in the dining room, as before. The 'engineers' cauldron' is no different to the 'second cauldron' now: a large amount of water and a little speck of cabbage, potato, flour, sometimes with traces of meat at the bottom. They serve us a second dish in the morning (what an achievement!), which is a wheat porridge (made of whole, uncooked grains). These grains pass through almost undigested, to the great happiness of sparrows and other birds. However, the poor quality of these dinners does not bother me now (certainly, for the time being). I usually boil a full pan of small potatoes in their jackets the day before they are needed, and clean the pot while hot. The next morning I put the cooked potatoes into a small bag (which I sewed from one leg of my old torn drawers), and I add this potato into all the porridge we are served, then

I add salt, some green onion, dill… After all these additives the porridge becomes completely edible.

The recent days were occupied with such culinary matters. I hardly think about my work… But don't worry about me. First, our chief, Rita Vulakh, expresses no interest in our work (malicious rumour-mongers are going about saying that she is having an affair with one 'Fritz'). Second, they are currently plastering the rooms in the quality control department, so the quality controllers do not have anywhere to work, and *a priori* it is understood that the quality controllers cannot work to the normal standard. Third, I have gradually shifted all of my responsibilities over to the temperature controller Sasha Lir. He goes all around the furnaces, writes the routine into the register, measures temperatures with an optical pyrometer, and also fills in the logs, marks the bricks (he writes the type and the consignment on the piles of brick in whitewash diluted with kerosene). Because of this I have a lot of free time, which I use for culinary purposes, or for example today to make my storage box. If only I could live together with you as a 'free' person, even at this brick plant, I'm sure it would be pleasant to make this box, to bring it home… But it is not interesting to labour only for myself, especially because there is no assurance as to what may happen tomorrow: they could move me to another barrack, to another plant, to another city. But I have decided to run a risk and to build up a reserve of food.

How are things with your garden, Lidusya? Now the weather is fine again. It would be good if you could harvest potato from the UBAS's fields before the rainy weather returns.

Incidentally, I have recalled two significant dates. Today, the 20th of September 1942, I have resided for exactly half a year at the Nizhni Tagil brick plant of NKVD *Tagilstroy*. Precisely at 9.00 p.m. on the 20th of March our party dragged itself with our belongings to the barracks and they placed us on the ground floor of the tenth barrack. At that time the future was completely unknown, and it was sketched out in only the very gloomiest of colours. The assistant commander of the column unequivocally stated that we should give up any hope of work in any speciality, and he anticipated that we would all be employed as general labour.

The second significant date is the 16th of September 1942, my 32nd birthday, and I offer my congratulations to you, my dear wife.

However, please don't think that I am offended that you forgot about this. With your mad life it is possible to forget more than such a trifle as the birthday of a husband. And it's senseless to remember such things. If only we lived together, it would have been possible to arrange a small family occasion on this day, regardless of the present situation. You would bake a 'friendly neighbours' cake, we would bake a vegetable pudding, would put some 'oil' on the table, would arrange a sumptuous feast. But now there is no reason to mark the occasion that I become one year older. On the contrary, I want to stay young in order to build a new life together with you based on new principles. I do not visualize clearly yet what this new life will consist of, but I think about it a lot. I will try to precisely define its content in the following letters.

Sending you a strong kiss, my dear, you and daughter.

Your 32-year-old — but still young — husband.

A few days ago Sasha Lir received warm clothes and boots from his mother (from Tyumen) through a friend.

Letter 29

Tuesday 22nd September 1942

My love! Eugene Mikhaylovich's departure was delayed by one more day, and I take this opportunity to add another pair of pages to my letter.

I was in the city today to seek treatment for a toothache. Yesterday I obtained a warrant; today I went to the city by car accompanied by Feldsher. I got off in the city centre, and Feldsher carried on further, leaving me to myself. A large hole was found in my tooth, so they have killed the nerve for the time being. The follow-up appointment will be on Thursday. Moreover, I will probably see a different doctor, since the woman who treated me today will have to go to work on the harvesting campaign. After my tooth was treated I made a small excursion to the centre of Nizhni Tagil. The only product I purchased was the only product in the entire store: un-ground and non-roasted coffee beans for 42 roubles/kg (half a ration of bread!). I sent a kilo to you, since I think that this product is seized on more rapidly in Sverdlovsk, and probably you have no coffee. I wanted to send you white felt boots, but unfortunately the shoemaker, who had promised to finish the felt boots today, said this morning that they will not be finished until tomorrow. So, it is necessary to await another opportunity. In fact, it is probably better this way, since Eugene Mikhaylovich will be travelling with his belongings, and such an extra load would be hard to carry.

Yesterday I went into the settlement in search of potatoes again. This time the exchange turned out to my disadvantage. I swapped a bucket of potato and a huge head of cabbage for three rations (2 kg of bread). True, the price of potato has begun to rise already; a bucket costs 170 roubles, a big head of cabbage costs 40 roubles, and a kilogram of bread costs 100 roubles, so that the exchange was close to the market prices.

Lidusenka, I have not received any letters from you for a long time… I have come to the conclusion that writing letters more frequently makes the task of writing easier. I have more than enough subject matter to discuss, and the letters turn out more interesting, since I can describe events spanning only the last two–three days of my life, when different details are fresh (I retain these details in my memory especially in order to share them in the letters). But when you write letters once in 15–20 days, then you describe changes in this period, you write only about the big changes, and the details are forgotten; but if there were no large changes, then I guess one will have difficulty in finding something to write about.

Thus, I hope I have convinced you, Lidusya. The only matter remaining is to find enough time to write these letters. But I am confident that you will find this time in the institute or at home if you will remember your husband more often. I finish here; kisses to daughter.

Your Armin.

Letter 30

Thursday 24th September 1942

My Dear!

So, Eugene Mikhaylovich has left. A happy person! Today he spent a happy day in Sverdlovsk among his relatives. Soon he will see his wife and son. Just yesterday in the evening we sat with him around this table, and I fed him before his departure with a friendly snack: '*kartoffel mit Hering*'. Here the evening has come once more, and again I sit at this table ... alone. But I believe that we will also be together soon. Can it really be true that we will not survive this war? Of course, there will be some tests we must withstand. At present I consider my life comparatively good and possibly I will look back upon it with envy.

I am now satisfied. I have eaten my fill. It is shameful for me that I write to you about my diet so much. It seems that a clever and talented person must not give so much time to the question of filling his stomach. Now I'm satisfied, but the problem of nourishment for the foreseeable future is the only thing I think about... Thinking about building up a certain reserve of potato. Potato! With what pleasure I eat it now in different forms. Especially boiled. We enjoy good living conditions in the room we have now: 1) stove and firewood; 2) electric heater and electric power. We are provided with firewood better than any other soldiers since the manager of the joinery and a dispatcher of the loading shop both live in our room; any unusable boards and other wooden materials surplus to requirements are sawn on the power saw, and the dispatcher of the loading shop made the order to deliver this fuel to the barracks. The electric heater is of a remarkable construction! It is based on the poor conductivity of water. Our neighbour, who is a driver, made it out of two porcelain bases obtained from some safety plugs and a food can. I have sent you a diagram of this heater so that you can make one. Two small metal cylinders from the can are pierced in several places, and are soldered with tin on the seam. Wires are then soldered to each of these small cylinders. It is essential to ensure that the wires are properly insulated from each other by at least 20–30 cm. This heater should then be placed on the bottom of a saucepan before it is filled with potato, then the water fills the remaining space and the heater is connected to electricity. After 15–20 minutes the water boils, and the potato is ready after a further 15–20 minutes.

It is possible to prepare cabbage and other vegetables in the same way. Do not add any salt (!), since in this case the conductivity rises too much, such that the current becomes too high and the plug will burn out. Tea, brewed in a cup, is ready in 12 minutes. For the above-mentioned purposes this heater is much better than a hot plate. But it is not applicable for the heating of soup (salt!) or porridge, of course. And it is better not to boil milk. I advise you to make this heater, since it will save you considerable time when boiling water for the morning tea or boiling vegetables for our daughter before she goes to bed. I will try to obtain such a heater and to send it to you at the first opportunity.

Every day I usually cook a large saucepan of potatoes with this heater, and I leave it until the morning. In the morning I put it in a small bag to be added into the soup in the dining room. That's how I get a 'thick skilly'. Recent days our menu in the dining room is scant in calories but diverse in content. Yesterday there was vegetable marrow instead of potato in the soup (!); today there was some fodder beet (!). The second dishes (twice daily) are even more diverse. Yesterday in the daytime five ruffs (salted and boiled) were provided; bream was given in the evening.[50] Today it was wheaten porridge at lunch and a salt-boiled pike (!) with a garnish of cucumbers in the evening. I add salt and the cold boiled potatoes aplenty in all of these dishes, since all dishes except the fish contain hardly any salt. Therefore I again thank you for the salt. Lidusya, it must surely seem ridiculous to you that I describe the menu in such detail. But after the monotonous meal skilly with the wheat and bread, and this disgusting feeling of hunger, which begins to suck ones insides out less than an hour after eating the skilly, it is so pleasant to have all this abundance of our Central Urals nature (actually it's scanty), together with the pleasure of feeling satiated…

For example, this evening after the 6.00 p.m. supper I boiled up a saucepan of peeled potatoes with the electric heater, skinned a bream and a pike, and gobbled them up with some sliced onion… But that is enough about food…

You will begin teaching in the institute in six days. But now you're fully occupied with your everyday problems: vegetable garden, firewood, getting food, preparation of winter and autumn clothing and footwear for yourself and for our daughter, washing, preparation of potatoes and vegetables for storage over the winter. Tell me some details of how you manage to do all of these things and take care of our daughter. I wonder whether I will see my daughter soon? Write to me in more detail about our

[50]Both are types of freshwater fish — ASK.

daughter, about her phrases, tricks, pranks. It is interesting how she deals with the realities of life, the hard wartime conditions, what is her attitude to the war, to the fascist aggressors. Lidusya, please memorize all of this and describe it to me in your letters. Indeed I am deprived of the possibility to follow the development of my daughter; and if I even see her … let's say in a year, then she will be completely another person than now at the age of four and a half years.

Goodbye for now.

Your Armin.

Letter 31

Tuesday 29th September 1942, 5.00 a.m.

Dearest Lidusya!

I have sent two letters to you already: one on the 17th of September and one on the 20th of September, and this is the third, and the last news from you arrived with Vulakh on the 16th of September. Your silence begins to worry me. Is everything OK with you? Are you and our daughter healthy? Might I receive a letter from you even today?

But today my mood is more than good after the 'celebration' of the birthday of Vladimir Fedorovich Ris. The celebrations would be complete if only another letter from you would arrive.

I will describe you how this birthday was celebrated. I now work on the night shift; it formally starts at 8.00 p.m. and finishes at 7.00 a.m. Yesterday, the 28th, a steady downpour fell throughout the entire day. After the night shift I slept, and then I darned my coat. Suddenly in the evening, around 7.00 p.m., Ris declared that he recalled that the 28th is his birthday, 35 years old (born in 1907); he is the oldest in our room, and I am the third in this seniority. Boris Viktorovich Rauschenbach was glad to hear that news, and he stated that he would stop all his office business and undertake some domestic work (they obtained and cooked products together with Ris). I asked permission to join in, so our work was in full swing. I provided half a cabbage to the boiling pot, plus some turnip, beet, onion and potato to make borscht. Rauschenbach undertook the production of fritters. At 8.00 p.m. I left for the night shift. On my way from the dining room to the loading office I slipped in the dirtiest place (you saw our mud in May of this year). However, even this hitch did not spoil my good mood. I issued a few passports to the railroad cars carrying bricks in the loading office, and then I checked that everything was quiet and left to go home 'for an hour or so'. I was kept busy with domestic matters for a while, and lay 'to snooze' for half an hour. I awoke some time after 1.00 a.m. Ris with Rauschenbach were still bustling around the comfortably heated plate, on which the borscht was almost ready and the fritters sizzled.

The table was clothed with a large sheet of screen paper. We sat down and started on the celebratory meal. I presented three large onions as a 'gift' for Ris. In conclusion, we wrote collective postcards to Ris's wife. Rauschenbach depicted himself in this postcard as a lady of the house, who tends to numerous guests consisting of … one guest, i.e. me. During the meal we started to recall how, where and in what a situation the last birthday of each of us was celebrated. I recalled that on the 16th of September 1941 (it was already the third month of the war) we arranged a family celebration. All our relatives sat around the table in our room: two grandmothers, Mum, Dad, sister Marina, Elsa, Vitya, Tosha with Misha and, on the left, Modestus with Irina. You baked a 'friendly neighbours' cake, there was a vinaigrette; it was so good and comfortable, we drank to a rapid end to the war, wished happiness to me… And here, what a 'happiness' came out.

But these happy moments are finished now. I brushed off an abundance of drying mud, which had clogged the sleeve of my coat, and then took leave ceremoniously from 'the owner' and 'the mistress'. I left … to work, moreover I arrived exactly at the start of the night meal at 2.00 a.m. (!) After some *balanda*[51] I had a talk with Vulakh in the office (she was the person on duty at the plant this night). I also met the chief engineer of the Hoffman furnace. Then we measured temperatures in the Burer furnace with Sasha Lir and … baked the potato, which I had had the foresight to bring from home. And finally I sit writing you this letter right here at the furnace.

---

[51] A weak Russian soup or snack.

Now it is dawn already, and the furnace is managed by the new shift of burners, so it's around 7.00 a.m. already. I have to finish this letter and go to eat the morning *balanda*, since my partner will be waiting for me in the dining room (engineers receive dual coupons, one for two people).

I'm now done. The new day begins, and then again the night and the shift.

I began to draw a picture for Elza with some coloured pencils. And I picked up my trousers, which had been repaired. These have turned out to be remarkably good trousers. Thanks to you, my dear, for your care.

Kiss you, my dear.

Your Armin.

Letter 32

Tuesday 6th October 1942

My dear wife!

I begin to worry about your silence ever more and more: is everything OK with you? The last news I received about you was brought to me by Vulakh on the 16th of September. Today it is 20 days since I last obtained a letter from you, whereas I am already writing the fifth letter to you in this period.

Our conversation on the telephone has to take place this evening. But I have such a foreboding that this conversation will not take place. First, in my ignorance I booked a conversation with you 'with notification of the address', assuming that I will give the telephone number of Modestus to the intercity trunk telephone station and they will 'inform' you when you should be near this telephone. But it turns out that, on the contrary, I must wait for you near to the telephone. In the second place, I booked the telephone call using the telephone coupon that was given to Ris, with his name on. Now I am afraid that they will report to you that someone called Ris awaits you in Nizhni Tagil near a telephone. You might think that this is an error or, vice versa, you could think that something bad has happened to me, and that I cannot call you myself. However, in case you have already made the last assumption, you will probably try to telephone in any case, but you will be worrying for nothing. It comes out badly either way.

Thus, I have prepared myself in case this conversation does not take place and have set about to write you this letter, especially because Rashenbakh's wife arrived, who leaves tomorrow into Bilimbay through Sverdlovsk, and she will post it locally.

All this time I have been thinking, trying to explain your silence. But it cannot be that you simply could not find time to write at least one postcard in these past 20 days. I consider this assumption improbable. My first explanation is the unpredictability of the mail. Perhaps, you wrote a letter, but it was mislaid along the road, either by our staff, or it got stuck in the military censorship. Second explanation: you are in great difficulty; you do not want to distress me and therefore decided not to write at all, until it is over.

But what trouble could this be? Our daughter is ill? You have fallen ill yourself? Lost your job? Moved out from the city? Your firewood has been stolen from the forest? Our daughter has been expelled from kindergarten?

No! I cannot believe these explanations. Rather I *do not want* to believe them, but my thoughts always return to this. I cannot believe that you would not share your problems with me. However, you did not keep silent when they did not give you dining rights in the UBAS, and when our daughter was ill. I was glad to share your happiness with you, and I shared your distress in your disappointments. This helped alleviate the artificial isolation from my family situation that I am forced to endure. I live with you through happiness and disappointments, excepting only that I cannot help you in any practical way. If this second explanation is correct, then you greatly offend me, Lidusya. I want to live one life with you. I want to feel myself together with our family.

I must establish connection with you soon. If our conversation on the telephone today does not take place, then I have one additional coupon (also with the name of Ris, damn it!), which I will use this evening to order a call to be put through to Modestus's telephone number on the day after tomorrow (at 10.00 p.m. on the 8th, Thursday). If they do not warn you, then I will talk to Modestus or someone else from the Kler family and I will try to find out if everything is OK with you. And if it is OK, then I will berate you on the telephone.

I assume that Modestus will 'apply pressure' on you, and that you will write a few words after all to your unloved (?) husband. Could it be true that you simply forgot to write me a letter for so long a time? Or perhaps you do not know what to write to me about?

In this case then I will help you with some questions of my own.

1. Firewood: Yours, Mother's, Golubovsky family's. Transportation of firewood from the forest. What are the prospects? The same about Tosha and Modestus.
2. Potato: Harvesting, at the UBAS field, drying, storage, amount; is it enough to last until the next autumn, and for seed (!); the same for Mum, Tosha, Modestus?
3. Other vegetables: Onions, cabbages, beets, turnips, carrots. Your work on the collective vegetable storage; price of vegetables on the market; the same for Mum and Tosha.
4. Classes in UPI: Arrival of students from the harvesting, new programmes, workload in hours and on what days of the week, comparison with other instructors: Lyustrova, Lipatov, Parysheva, Kochin; scientific research, Kikoin.
5. Nourishment: What products did you buy in September with the ration cards and what else besides the cards? Change in the prices during October. Dining room rights for October. Quality of dinners.
6. Daughter — this question should be placed first — her health, kindergarten, behaviour, figures, phrases, does she recall anything about her dad, eating at home, milk, oil, fats?
7. Your life, health, mood, behaviour (!), entertainment, fulfilment, clothes.
8. Clothes for autumn and winter for our daughter: coat, boats overshoes, felt boots, breeches, warm dresses. Your shoes, overshoes, warm dress and other things. What can be obtained on the goods ration cards, what did you purchase in August and September?
9. Home life, evening illumination, electrical heating, firewood for the stove in the kitchen; mutual relations with the mum and Golubovsky family: storage of potato, the cooking of food, cleaning the restroom, water transportation.
10. Golubovsky family: how many men, home concerts, our piano.

11. Relatives: Their health, in particular the health of Mum and Granny Olga; the arrival of Marina from the kolkhoz; her participation in domestic life (washing dishes, baking bread); Grandfather Sasha's workload, his violin playing; Mum's work in the institute, cheerfulness, old age, concerns (whom from the family does she helps with money and troubles at the moment?). Tosha: her life, children, Nyura, apartment, firewood, etc. Relatives from Krasnoufimsk — Appolinaria Fedorovna,[52] Lyusya,[53] Valentine[54] and their life; Volodya and Pete;[55] Modestus, Irina.
12. UBAS: the people at UBAS, and so forth.[56]

Lidusya, if you would make each of these questions to be a theme for one letter, then it would make 11 large letters.

I will write about myself in the following letter, which I will write to you on the 10th, since I decided to write you a letter every five days. I live a completely calm and regular life, almost without any concerns, but with some indeterminate expectation of change. My enthusiasm for collecting potatoes has passed away gradually, and I begin to once more be occupied by sciences and to read fiction.

Kisses to you. Forgotten by you, HUSBAND.

Letter 33

Saturday 17th October 1942

My dearest Lidusya!

The day before yesterday I began to write you a letter about the fact that I had not received one letter from you. However, yesterday, the 16th, the mail was given out, so I received four letters: a letter from Marina (dated the 3rd of October), two postcards from you (dated the 5th and 9th of

[52] Lydia's mom — EAZ.
[53] Lydia's sister — EAZ.
[54] Lydia's sister-in-law — EAZ.
[55] Lydia's brothers fighting on the front — EAZ.
[56] An additional afterthought written in small letters — EAZ.

October) and a postcard from Karpachev (dated the 7th of October).[57] Clearly the censorship has begun to detain our letters for a very long time. I was struck by the pessimism of your postcards, Lidusya. You are in a complete whirlwind of emotions. On one hand there is your work at the institute; on the other, the cursed everyday questions: kindergarten, firewood, nourishment. It seems to me that this explains your phrase about the 'hostile attitude' of our relatives. It is a well-known fact that a daughter-in-law and a mother-in-law often do not get along with each other. But my Mum is one of the best mothers-in-law. She barely stays at home, she is always willing to help everyone; and I am confident that she is also good to you (Marina writes that they sewed a dress for Elza from mother's sarafan). Please forgive Mum if she likes to gossip with some friend. After all, she's quite old, so it's harder for her than for you. What a crowd of dependants she has to feed; and with everyone claiming personal requirements too (Nyura, Granny Olga). Evidently, you have experienced some conflicts with Nyura and Grandma Olga concerning your everyday matters. But these trifles, of course, arise all the time because of the shortage of everything. Moreover, these shortages and hardships put everyone in a bad mood, such that each petty squabble could easily grow into a larger spat and easily cause lingering hostility.

Darling Lidusya, I beg you to try to get on with Mum somehow during these hard times. Please help her at any difficult moments with words and deeds (I have no doubt you do this already). She has had a life full of trials and tribulations, and I'm afraid in case these past few years will destroy her completely. By and large, Lidusya, I'm sure that you manage to settle all these everyday conflicts because of your tact and intellect.

I only regret that some conflicts appear to have arisen in the first place. But, apparently, it is inevitable during this difficult time. These past few days I have been thinking a lot about our future family life. We will definitely live apart from any other relatives. It is best to leave for another city. My basic goal will be our family's prosperity — good food and no shortage of anything. True, Lidusya, I am thoroughly gripped by 'the science bug', so this will be a serious obstacle on the way to realizing my plans for 'family happiness' outlined above. But I think that it will be

---

[57]The scientific chief of AGS — EAZ.

possible to find a compromise solution. In the beginning, I will use my science to gain material prosperity. Dreams, dreams…

However, Lidusya, we should not divide up the bearskin before the bear is shot. The immediate task is for us to survive these present difficult times as painlessly as possible. Nevertheless I think that Comrade Stalin's order will be executed and this winter will be the last military winter. So let's brace ourselves. Let us not turn our attention to the various, annoying, everyday trifles. But if such daily frictions do appear, then we will try to settle them as tactfully as possible, without allowing the spark of non-communication to flare up into the flame of quarrel or hostility. Dearest Lidusya, I firmly ask you to tell me about all of these everyday annoyances, and how you manage to settle them. I don't need any cheerful letters, since my mood is completely cheerful already. The only thing I want is to be together with you and to bear the rough with the smooth, together, with you. But your silence separates your life from mine. It is understandable that you are in a poor humour. But why do you punish me because of this? You force me to feel lonely and torn from my family. In order to support me it is completely unnecessary to write cheerful letters. The only thing I want is to know all details of your life, to know how you cope with any difficulties. I want to help you with advice and action (unfortunately, I can only do the latter within my narrow confines of ability). But you do not write to me about anything. For example, I learned from the letter of Marina, that your sister Lucy arrived from Krasnoufimsk, that a new dress was sewn for our daughter… Marina wrote to me a series of phrases and facts from the life of Elza. Lidusenka, I hope that you understand me, and that you did not take offence at my moralizing-philosophical letter. I also think that our correspondence will subsequently become more intensive.

This is the third day of winter in Nizhni Tagil. The temperature is 3–5°C. Everything is covered with snow. It seems that winter has arrived permanently and in earnest. But I continue to walk in the autumnal style: peak cap, coat and rubber boots. Actually, I put on a sweater under my coat to keep warm, and woollen socks on my feet. The cobbler has finished the first pair of felt boots. I will bring them to Lisa[58] the next day to

[58]A friend in Nizhni Tagil — EAZ.

store until an opportunity to send them to you or a suitable chance to sell them in Nizhni Tagil arrives (they are a bit small for me). My felt boots will be also finished soon and then I can get them.

I live a calm (comparatively) and regular life. This week I was on the daytime shift. Sasha Lir needed to fix his boots and was not at work for four days. This coincided exactly with the need to analyse the furnaces' routines for the second half of October. This work somewhat enticed and distracted me from any culinary occupations. I cut out boards of plywood, with a red star in the corner (also made of plywood).

Every morning at 7.30 a.m. I go to work with eight to ten potatoes in the pockets of my coat, which I bake in one of the three calcining furnaces I am in charge of. For this purpose, there is a twisted copper wire in the inside pocket of my coat which I use to string potatoes and put them into the fuel inlets of the furnace. At first my potatoes were half-cooked or were heavily charred. Now they are baked with a slightly browned crust. It's a pity that I do not have the chance to entertain you with this potato meal.

For the fourth day in a row they have given us a thick skilly made of ground cereal in the dining room. It seems so tasty after the skilly made of water with some non-boiled wheat grains at the bottom (they fed us with this for the last month). But, in truth, I did not like this ground cereal until now.

I continue to exchange bread for potatoes. Over the past week I have accumulated two and a half rations of bread (2 kg) and bartered them with the same woman for one bucket of potatoes (1.6 kg of bread); and she gave me the head of a cabbage, three beets, eight carrots and one turnip for 400 g of bread (equivalent to 40 roubles). Using the hot plate in our room, I cook conserves made of all these vegetables, or I eat them raw.

I have temporarily dropped my studies 'in science' (*Articles on the Physics of the Atom*)… 'Housework' takes up all my time!

It's warm in our room, since there is plenty of lumber of many kinds at the plant for the taking, which the manager of joinery allow us to take (he lives in our room) and which was transported here using a horse from the loading shop dispatcher (who also lives in our room). Because of this combination we have an abundance of dry firewood without any problems, and it's warm, although the second window frames are not inserted yet.

It is so hard for me to read (sitting in the warm room) that we do not have firewood at home, it's cold and our daughter shivers at night, you hands freeze while you write your letters… I would sooner exchange the warm but cheerless room here for our cold room at Chapaev Street.

Indeed, no cold is so terrible if we get together under one blanket (and throw that motley Persian carpet on top) and snuggle up to each other...

Your HUSBAND.

Letter 34

Monday 26th October 1942

Dear Lidusya!

Today your postcard arrived dated the 16th of October, and the letter from Marina dated the same day (after entering censorship it was passed to Sverdlovsk and was sent on again on the 18th, arriving in Nizhni Tagil on the 21st, and was received today on the 26th). Nonetheless, these letters take so long to be delivered from you to me, a whole ten days (probably they arrive faster in return). Still, I spoke with you on the 21st, and I know the content of your postcard already. On Tuesday the 3rd of November I will try to phone you again.

I know from our telephone conversation that our daughter is recovering. I hope that now she is already returned to full health, and your headaches are gone. The lack of firewood in our apartment distresses me most of all. I saw driver Schmidt today. He recommends turning to his friend Leonid Tretyakov through Dusya Isaeva to arrange delivery of the firewood. This driver lives somewhere in the suburbs of the city, but he happens to visit Schmidt frequently, so it will be easy for you to get in touch with him through Dusya Isaeva. Probably you could arrange to transport firewood from the timber yard. If he is not able to help, then he might recommend someone else. It is important to find a contact amongst the drivers.

It is annoying that Golubovsky's arrival was delayed due to his wife being sick with typhoid fever. I hope that they have already arrived, and started heating the stove in the main room.

Write to me in detail about what vegetables you have stored up for the winter and how you have preserved them. You said on the telephone that the brine leaked out from the pickled cabbage. Probably there is a hole in the barrel? It's a pity you are so far away; one soldier, an assistant in the physics department of Sverdlovsk State University — Pavel Ferdinandovich Kromer — makes the most remarkable barrels and exchanges them for two buckets of potatoes each. Earlier in his life he helped his father in the Caucasus to make vats for grape juice.

This past week I have worked on the night shift. The office of the quality control department was refurbished; we light a stove at the beginning of each evening by taking lumber found around the area of the plant, so that the office of the quality control department has become quite comfortable now. I studied the *Articles on the Physics of the Atom* until 10.00–11.00 p.m. (I have once more started to read this book after almost a month of 'vegetating'). Between 11.00 p.m. and 1.00 a.m. I made my rounds of the furnaces to check that they were running according to the established regime. Between 1.00 a.m. and 2.00 a.m. we went with Sasha Lir 'to hunt' (without a coupon) for some night skilly and were successful almost every night. Each time it was necessary to think up a new story to explain the lack of coupons for the night dinner (because if we take a coupon for the night dinner then we can not eat the daytime dinner). However, the cooks on duty have already become accustomed to us, so they give us the skilly without a coupon and almost without objection.

After this we arranged a 'dead hour' in the quality control dept. lasting for several hours; and in the morning we made up sums for the shift, filled out the logbooks, drawings, etc. At 8.00 a.m. I went into the dining room, where my partner, Fridrikhsen, who worked on the day shift 9.00 a.m. to 12.00 a.m. arrived with a coupon (the double coupon for himself and for me). I caught up on lost sleep after my night shift (in addition to the night's 'dead hour'). At noon I went to the dining room for dinner. Then I returned home and was occupied with my business: preparing some borscht or potatoes and darning some socks.

I wonder why darning has begun to take so much time lately, since my threadbare trousers are bulging at the seams. But every cloud has a silver lining.[59] I have now learned how to properly darn seams, to sew patches 'through the edges', to make buttonholes, etc.

I read fiction now and then. I read the *New Tablet* by Panteleimon Romanov, *The Twilights* by Przybyszewski, *The Good Soldier Švejk, Volume II* by Hašek. *Švejk* was especially pleasing to me — it is wonderful. This is an ingenious parody of the war.

They took away the radio loudspeaker from our room (it is not allowed!); I barely read newspapers, so that it is impossible for me now to follow the course of events in the same detail as it was earlier. The matter concerning the creation of an international tribunal and about the demand of the law courts to remain above Rudolf Hess is vigorously discussed.

The skilly problem is probably now about to improve, because a shipment of new oats — ground cereal — has arrived. Breakfast is a soup of ground cereal, cabbage and some signs of meat; dinner — porridge (!) of ground cereal — that is good but too little; supper — soup made of flour and yet more ground cereal. With what pleasure I now eat this ground cereal, recalling with surprise (!) that earlier I did not like this porridge very much — life determines our awareness.

I'll finish here. It is time to go to the furnace to prepare some baked potato according to the method of No. 1, and then to eat the evening skilly.

I await a letter from you with the answers to the numerous questions I asked in my previous letters. Strongly kisses to you, my darling.

Your Armin.

Letter 35

Sunday 8th November 1942:
The anniversary of the October Revolution!

My dear wife!

Today is another big date that we used to celebrate together so gaily and so solemnly in previous years.

---

[59]Originally expressed in Russian as 'there is no harm without some good' — ASK.

It is winter. Minus 20°C, in the office of the quality control department, 2.00 a.m., Sunday the 8th of November. I'm on the night shift. And this day is no different to any others. Even the skilly is the same.

Winter has now truly arrived. Yet only three days ago Lir and I returned on the evening of the 5th from a field about 10 km from the plant, with two buckets of potatoes each, which we harvested in the potato field of the state farm (earlier the field had been harvested by a tractor). The weather was warm, and we delighted in taking in the autumnal view and walking in the forest alongside a road. But an hour after our return to the barracks the snow began to fall, the blizzard began, and the real winter came on the morning of the 6th of November.

So it was impossible for us to complete a third hike in search of potatoes. That is a pity. We would bring not less than three buckets of potatoes each, since on the 5th we found a rich layer of potatoes, but there was no time to 'exhaust' it.

However, I will describe, in order, our potato marches, since they were a pleasant and useful variation in our otherwise unpleasant, useless and monotonous life. I already told you on the telephone about our first trip on the 2nd of November. It was hardly successful. Lir and I had one shovel between the two of us, and without having any experience in this activity we repeatedly dug in places that we had dug over already, so we barely harvested half a bucket of potatoes each.

On the 5th we left for a second time, which proved to be more successful, although at first the prospects were poor. First of all we could not find any shovels at all, since all of them were pilfered from the plant by an enterprising 'Fritz', and were hidden in the bushes next to the field 'for the next time'. After reaching the field (8 km from the plant) we first started to search for the shovels in the bushes. After an hour of fruitless searching I luckily struck upon two excellent shovels in the bushes. Our mood improved immediately. However, it must be noted that the 'good mood' of Sasha Lir is a relative concept, since he is a terrible pessimist and a melancholic. We harvested potatoes for a while in the large field accompanied by numerous 'Fritz'; but we were convinced that 'the layer' [of potatoes] was already exhausted by previous 'potato-finders'. Then we shouldered the shovels and left with Lir 'to search for happiness', walking briskly further along the road. We walked and walked ... and had just decided to turn back, since there was no sign of potato fields any more, when we were lucky again, since a young person who guards hay in the state farm's fields appeared on the road and agreed to be our conductor. After walking about two kilometres along the road, we turned right into the forest on his direction. Walking through a swamp for approximately 100–200 steps without any path we suddenly stumbled into a small potato field — 1–2 hectares — surrounded by forest, where the foot of 'Fritz' had never stepped. Of course, this field had already been harvested by tractor, but yet it had not been dug over by anyone. We set to work enthusiastically; and our enthusiasm flared up ever more and more as each shovel picked out large pink potatoes from the ground. After every 10–15 minutes of digging I had a break, selected the largest potato from the freshly harvested pile of 10–20 potatoes, peeled it and ate it uncooked. I never thought that raw potato could be so tasty. I ate 10–12 raw potatoes that day and without any ill-effects (!) — not even a stomach upset! That's what enthusiasm for work means!

But after two–three hours we had to interrupt our work, which was in full swing, since night was drawing on. As a result we harvested one and a half to two buckets of potatoes each, hid the shovels in the bushes 'for the next time' (alas, this 'next time' will have to be put off until the next autumn); and we left for home with our 'prize catch', eagerly discussing a repeat of the trip the following morning (after the night duty). But winter did not wait for us, and the following morning all the fields were covered with a thick layer of snow. Altogether we harvested two buckets of potatoes each in two trips. These buckets of potato are equivalent to 3.2 kg of bread = 5.3 rations of 600 g. This is good, but not good enough. Some 'Fritz' had time to make four or five campaigns (initially in an illegal fashion, since the official permission for 'release to dig for potatoes' was only given on the 1st of this month) and they collected three to four buckets of potatoes every trip. Potato, potato, potato... This potato fever has gradually overcome the entire force, and in recent days hundreds of soldiers from the force (these are certainly off-duty workers) have left to go harvesting. The only topic of conversation is about how much has been harvested.

But that is enough talk about potatoes. Excuse me, Lidusenka, that I have given over so much space in this letter to our potato problems. But these marches were such an unexpected variation for me. It was such an unusual feeling to me that I was able to walk freely in a forest (for the first time in half a year of my confinement in the construction crew I was in a forest)... And on the other hand, this was a definite solution to our problems of finding food, which will be on the agenda repeatedly over this winter...

Yesterday evening (the 6th) Schmidt told us with regret that his wife had gone back home already. It's a pity that I did not send the felt boots with her. Now I have to wait until the next opportunity, or exchange them here in an hour of need...

I live as before in all other respects. I was on the night shift this week (because of this I had the chance to walk in search of potatoes during the daytime). From Monday the 9th I will work on the day shift for a week. On the 10th I will barter 800 g of bread for 1 kg of onions (36–40 onions), and 400 g of bread for different vegetables. Onion is my major exchange fund. I acquired four buttons for the padded trousers from tailor Fuchs for two onions. He will make me ear muffs for my lambskin cap for ten onions (these ear muffs will be folded inward in the warm weather). I promised to give a further 15 onions for the resoling of the second pair of felt boots. I want to try to obtain some more salt in exchange for the onions etc. Lidusenka, as you see, your husband begins to show some talent as a 'businessman'. By the way, I'm running out of salt already. If opportunity arises, please send me 200–300 g of salt, white writing paper and some different buttons.

How do you live without your husband, Lidusenka? You write little to me about your life, work; about our daughter. I know that you do not have time, that it is dark at home after 6.00 p.m., but nevertheless you could find some time in the institute once a week during a break between classes, maybe during a free period. It is good that we succeed in supporting each other on the telephone. I thought much after our last conversation. Will they really mobilize you as well to work, and will they send you out somewhere else from Sverdlovsk? This will be a very difficult for us. Write to me in detail concerning the mobilization of Postovsky's wife. Where is she now? Where is her child? Where is Isaac Yakovlevich?[60]

---

[60]An organic chemist from the UII — EAZ.

On Tuesday the 17th at 10.00 p.m. I will ring you at the telephone number 7859; on Sunday evening (the 15th) I will order this conversation in advance; tomorrow (the 9th) Rikert will purchase the telephone coupons for me in the city.

I'll end here. I await any letters from you. Ask Marina and Mum to write to me about Elza. I always read about Elza's different phrases and her tricks with great pleasure; and then I talk about these to my companions-in-misfortune.

Kisses to you all, my darling. Regards to Modestus and Irina.

Your Armin.

Letter 37[61]

[Letter 36 is lost.]

Dear Lidiya!

So, our telephone conversation on the 17th and again on the 19th did not happen. No one even approached the telephone on the 17th, and I was waiting for two hours nearby. On the 19th they connected me with Modestus after an hour, but the line was very bad. Modestus could hear me only occasionally. I could barely distinguish anything of what Modestus said because of these periodic interruptions three to four times per second. Nevertheless I understood that he had alerted you, but that you did not come for some reason; that you are healthy, but frequent headaches torment you.

So I have not received a letter from you for more than half a month. Schmidt gave your last letter to me. Apparently, I should accept your reasons and not wait any longer. However, I was told about your recent days from the letters of Marina and Mum, therefore I am not angry with you.

Lidusya, I want to write to you about my economic successes and … failures in the past 20 days of November. On the 2nd and 5th of

[61] This letter consists of one page folded into a triangle with the address and stamp — EAZ.

November I made trips with Sasha Lir to a state farm in search of potatoes (I wrote to you about this already). On the 6th there was a heavy fall of snow, the temperature plummeted below zero, and so we had to end these potato-gathering trips. We collected two buckets each. I was on duty in the daytime from the 8th to the 15th. Moreover, towards the end of week I noticed that my reserve of potatoes was coming to an end. It was therefore necessary to give serious thought to my existence in the immediate future. I came to the conclusion that I would not live for very long on 600 g of bread plus three liquid bowls of skilly a day. After coming to this conclusion I decided once again to build up a certain reserve of potatoes at any cost. And, already, I now have five buckets of potatoes. But it was many times more difficult to obtain five buckets of potatoes in November than it was in September. Then, people sold off and ate potatoes; now, they don't want to sell any more. This has had an immediate effect on the price, which, for a bucket rose from 150 roubles then to 250 roubles now. Accordingly the price of bread has risen from 100 to 150–170 roubles for 1 kg. Up until the 1st of November I exchanged bread for potatoes with the same woman. After the 1st this was no longer possible. Then I exchanged pieces of laundry soap — 300 g and 150 g — for one and a half buckets of potatoes. A 400 g piece of laundry soap costs 240–280 roubles on the market. But this it is not enough. Then I visited the collective farm market three times at my own risk and with real fear, and sold my new padded trousers for 450 roubles (I repaired and wear the ones you purchased for me for 60 roubles) and a new set of linen for 200 roubles. I purchased a 370 g piece of laundry soap for these 300 roubles, and a loaf of bread — about 1.2 kg — for 180 roubles. I bartered the soap in the settlement for a bucket of potatoes and the bread for three-quarters of a bucket of potatoes. Furthermore, a certain arrangement with some liquid soap became possible. I bartered 700 g of this soap for a bucket of potatoes; moreover, half a bucket was obtained as the 'commission'. Altogether with the previous reserve of one and a half buckets I now have about five buckets of potato. I hope that this reserve will be enough to last me up until the New Year. But we shall see when the time comes. Up until now, despite three attempts, I have not been able to sell the spare pair of felt boots. Previously I was offered 1400 roubles for them, but it seemed too cheap. Today they would only offer 800–1000 roubles. I left the felt boots at Lisa's house. I plan to sell them after the 22nd at any price I can get. Next I purchased some fats (oil and fat) for 500 roubles. If I am able to sell the felt boots, then I will live magnificently up until the New Year. Lidusya, I took your shoes in to be repaired. They will be finished soon.

Your husband-trader Armin.

[Notes in the margin]

I finished this letter for technical reasons. Wait for a continuation. Kisses to you and daughter. Wait for a telephone call on the 1st of December at 10.00 p.m.

Letter 38[62]

Thursday 3rd December 1942, 4.30 p.m.

My dear Lidusya!

At long last I had a chance to talk with you on Tuesday the 1st of December. After each such conversation I remain in a good mood. Moreover, I am left with the same impressions and feelings as when you visited me, and I talked with you about all our domestic matters, not on the telephone, but directly in person. Wherever I am these days my mind travels back over our conversation about your life, about our daughter. I am very glad that your headaches are not so strong now. It is good that you managed to pay off your debts. I don't understand what 'fur' coat you sold for 1500 roubles. Was that your suit coat? This is not good. If you said it was one of my coats — that would be better.

[62]One page is folded into a triangle with the address and stamp: Sverdlovsk city, Chapaev Street 8, Flat 1. To Lydia Mikhaylovna Stromberg.

Lidusya, I am sure that you are angry with me for selling 'Tosha's felt boots' so cheaply (750 roubles), but the fact is that these were not felt boots, but one continuous misunderstanding. It's a result of an unsuccessful '*geschäft*'. I will describe the situation to you frankly. After your departure several days passed, but the man selling the boots still did not bring them with him for different and increasingly extraordinary reasons. During the second week Fridrikhsen went on to the day shift, and I worked on the night shift, and thus I had to solve this problem by myself in this exceptional circumstance. Finally, the felt boots were brought. I inspected them to see if they fitted. These white felt boots were utterly without a sole, with very shabby uppers, and were a size too small, so that my toes jutted out when I put them on. I decided to keep them for one more day in order to discuss the possibility of repairing them with the shoemaker. But their sly owner saw that I was undecided, so he asked for two cups of tobacco for the boots (which I had exchanged for a bundle of light tobacco previously), and would not agree to leave the boots free of charge in any event. And here I made my biggest mistake, for which I have scolded myself a hundred times already. I gave him one cup of tobacco, and this 'burned all the bridges', since after this only two options remained — to take the felt boots or to lose the cup of tobacco (200 roubles!). Another pair of felt boots appeared the next day. These were considerably better: hemmed, of a large size, with thick soles, but with only one defect — one of the boots is large and black in colour, the other is a bit smaller and red. The vendor requested one bundle of light tobacco for the second boots. I decided that these felt boots would surely be warm, and bought them. I asked Brousman to sole the first pair of felt boots so that they could be sold in the winter for a profit. But they turned out to be of very poor condition. I was utterly convinced of this when I tried to sell them at the market. During the four days in which I unsuccessfully attempted to sell them, I heard the same remarks dozens of times: 'They're a pointless use of leather', 'They won't be warm', 'The patch on the front completely spoils the finish', 'They need repairing', and after such critical observations they proposed 800 roubles or generally nothing, and returned them back to the ill-fated salesman, i.e. to me. Finally, at the end of the fourth day when I had despaired of ever selling them and intended to go home, a peasant woman offered me 800 roubles for the pair. At first I would not agree to sell them for less than 1000 roubles, but seeing her moving away, I broke down and agreed. When she came to pay, she could only find 750 roubles. But I was so sick of these felt boots that I agreed and sold them for 750 roubles. That's how this felt boot epic ended. Please do not be angry that I sold them. It would be shameful to offer such boots to Antonina Mikhaylovna.[63]

[Added in the margin]

To be continued in the following letter. Your husband-trader.

Letter 39

Tuesday 8th December 1942

My dear wife!

I wrote you a letter on the 3rd after our telephone conversation on the 1st, with a description of 'the story of a pair of felt boots'. I now feel uncomfortable that you promised these felt boots to Antonina Mikhaylovna, but I did not give them to her, and instead used them for the purpose of my self-subsistence. But these felt boots were terribly scabby, which sets my mind at rest, and the 750 roubles that I obtained for them is not a bad price. At the same time I have recalled one amusing episode which occurred on the same day that I sold the boots. I decided to sell a box of cigarettes as well, which you had sent to me. After learning the price of cigarettes — 70–80 roubles per box — I started to sell cigarettes. A boy rushed up to me: 'How much are the cigarettes?' I answered: 'Seventy roubles.' He proposed: 'I'll give you fifty roubles.' Finally, I bargained with him and sold them for 60 roubles. To my surprise, the boy took an enormous sum of money from his pocket and coolly counted off 60 roubles for me. After ten minutes he came back to me again: 'Uncle, do you have any more of these

[63] 'Tosha' — Lydia's sister — EAZ.

cigarettes?' I asked what price he had resold these cigarettes for. The boy gaily answered: 'For 100 roubles, thanks to you I have earned 40 roubles!' Such a smart fellow! This incident amused me so I was not offended.

I have not received any letters from you for so long. You began writing your last letter on the 8th of November and finished on the 18th. But I am becoming accustomed to this, and the long silences do not offend me. It only becomes a little annoying when my colleague and roommate K.G. Fridrikhsen receives a bundle of three or four letters from his wife, written at one to two day intervals. Possibly this is because they do not have children and the wife misses her husband very much.

I enclose a picture: 'Elza cooks supper on a small stove for herself and Mum.' I think that she will be pleased with this picture. Please put it near her bed so that she will often think about her dad, looking at the picture. Please send me any pictures of Elza, especially those that she has sketched specially for me.

My position (work, way of life, food) remains unchanged, and I only hope that it does not change for the worse.

Sasha Lir had some bad luck. A new staff arrangement has come into force, and the post of temperature controller was reduced, so that today they even wanted to remove Sasha from the food allowance (not to give him any bread and skilly at all). It was only because I acted just in the nick of time that it was possible to attain an 'allowance' for him for one to two days. There are several vacant posts for him already: electrician (!), motor mechanic (!!). They are not within his speciality one little bit (!), but it does not matter. What is important is that this will ensure he gets 800 g of bread (!) daily, and the work is light.

On the 5th of December I performed at the amateurs' night in the brick plant's club as a pianist in a jazz ensemble. I practised my part a great deal with the jazz group for three to four days beforehand. In general, I think the performance went down well. I'll leave it here. If I don't receive any letters from you before the 15th, then I hope to learn about your life from our telephone conversation on Tuesday the 15th.

Your Husband.

Letter 41[64]

[Letter 40 is lost.]

Friday 18th December 1942, 11.00 p.m.

My darling!

There's one hour left before the 19th of December, and after 12 more days the New Year, 1943, will begin. New Year! What will it bring to us? We will hope that 'the new year will be better than the old one', as Eugenia Samoylovna said at the New Year's Eve party in 1940, 1941 and 1942. I am at least almost convinced that the New Year will be better. But in two days time I will mark three-quarters of a year = nine months of my stay in unforgettable Nizhni Tagil. Time passes more quickly now. This is good. It means that we will be together sooner. I will see my daughter sooner; will restart my research work; I will once again read foreign periodicals in the comfortable reading hall of the UBAS: the *Journal of Chemical Physics* in a blue cover, the large-sized *Zeitschrift für Electrokhemie* in the red cover... I miss home and my old job so much!

The pointless pastimes in here wear me out! However, the time that passes is not entirely pointless for me. The main advantage is that here I have encountered real life and real work. I have picked up a few useful habits for the future (if that would ever be!!): the darning of linen, cookery and even resourcefulness. For example, several days ago I had 660 roubles, and today I have 840 roubles, and have managed to keep all my belongings, and I still eat a 600 g ration of bread each day. The increase in my capital by 180 roubles was entirely due to a successful deal. I will write about the details another time. The important point is that bread costs one and a half times more on the market than it does in the barracks. I think to use any profit to purchase bread again, and then to exchange the bread for potatoes in the settlement: one bucket for 1.6 kg of bread [...][65] And one bucket of potatoes will by and large come out of the surplus.

[64]One large sheet, folded into a triangle with the address and the stamp. Marked: 'Examined by military censorship. Sverdlovsk'.

[65]The rest of this line was struck out by the censor — EAZ.

I can assure you that this combination was not easy; moreover, both the idea and its practical implementation can be assigned to my new commercial abilities. These are the matters with which I was occupied during this week of night shifts. From the 20th to the 27th of December it is necessary to get on with my work again. By the way, Lidusya, did you roast the grains of coffee that I sent to you with Eugene Mikhaylovich Schumacher? 'By the way' — because yesterday evening I very successfully roasted about 300 g of these grains in the laboratory, by placing them inside a square, iron cube (embedded into the furnace) that is used to determine the ability of bricks to adsorb water at its boiling point. Of course, the cube was empty, and we heated it in order to increase the room temperature from 4°C to 24°C. I placed the grains into the space between two grids used for sieving sand, fixed the grids with a wire, sat near the edge of the cube and shook these grains inside the cube, until they became dark brown in colour. The authorship of this process belongs to our laboratory assistant P.E. Rikert. But as the saying goes, I surpassed the author, and by his own acknowledgement I 'roast grains to perfection'. Then Sasha Lir ground the grains in a porcelain mortar. We boiled some water in a clay pot (part of some laboratory equipment made by our master potter Reynveter) with the electro-boiler. We poured the coffee into the water, and we had a collective coffee party the four of us together: Lir, Fridrikhsen, Rikert and I (the night shift of the quality control department in full strength). This is how I live, Lidusya! Time slips by with our work and our worries. And let this hard time fly faster for us both. It's especially hard for you, my darling. I can imagine that you have so many problems! You have to rush from home to work, from work to the dining room, from the dining room back home. A pile of the things to do: to feed your daughter, do the washing, darning, to prepare for university classes, to check your student's home work. And it is cold and dark! Brrr!

These past days I have been impressed by our telephone conversation. These conversations mean so much to me. Each time I am left with the impression that you have visited me, and I do not feel myself to be so lonely. Tomorrow Rikert should receive a parcel from his wife in Sverdlovsk, which was accepted (with difficulty!) with a warrant that Rikert sent from here. I am ashamed that I asked you for a parcel, since I have spare potato now. Please send something edible, which is easier (?) to find in Sverdlovsk.

But please do not reduce the necessary amount of rationed products for yourself and Elza. I end here. It's morning already. Kisses to you.

Your husband.

[Written in the margins:]

Maybe it will be possible to send some shoes to you through Lisa Chernobrovkina. I visited her at work recently. She offered to send the shoes to you via the cleaner, who drives into Sverdlovsk frequently.

I am very concerned about our daughter's constant illnesses — first her eyes, then the scrofula. Poor girl![66]

Please, make up several words from me to her. I have already started to sketch a picture for her.

[An insert of one-eighth of a sheet of paper into the letter:]

I enclose a permission slip to allow parcels to be sent to me. In truth, it must be mentioned that I don't know of any single case of a parcel arriving from Sverdlovsk to anyone. Moreover, I know that even with the existence of a similar official paper, packages were not accepted from the Sverdlovsk region. In other places parcels have been accepted with this permission paper. All in all, let's try. Do not forget to send SALT.

Letter 42[67]

Wednesday 23rd December 1942

My dear Lidusya!

I wish you a Happy New Year!

We bring in the New Year together but with little joy this time.

I have unwittingly recalled the company and the comfortable apartments of Eugenia Samoylovna, with whom we have celebrated the New Year so many times in previous years.

---

[66]Historically known as 'the King's evil', scrofula is any of a variety of skin diseases, particularly a form of tuberculosis, affecting the lymph nodes of the neck.

[67]A postcard with two stamps, the address and 'Examined by Military Censorship. Sverdlovsk'.

Lidusenka, what shall I wish for you in 1943?

1. The main thing is definitely an end to this war, so that we can once again begin to build our lives together under peaceful conditions.
2. I wish for our dear daughter, Elzochka, to grow up plump and healthy and smart.
3. I wish all our material difficulties would end, to be able to eat as much and of whatever the heart (and stomach) desires, to have all the necessary things we need in life such as firewood etc.
4. To live together with you, a completely happy life.
   Loving you — your husband.

[Written in the margins:]

If only half of these wishes would come true!

Letter 43[68]

Saturday 26th December 1942

Dear Lidusya!

More than a month has passed since I have received any letters from you (!!!). Your last letter was begun on the 8th of November and finished on the 18th. Naturally, I'm very distressed by this. In December I have only received two letters from Marina (dated the 1st and 12th of December) and a postcard from our daughter with your addition (dated the 16th of December). I console myself with the fact that I have information about your life in December 1942 from our last telephone conversation on the 15th. However, concerning the letters, I have convinced myself that you must have written a letter, but it was mislaid somewhere by our postal force. For example, some letters to me have definitely been misplaced. It has been said to me that a letter of mine was seen in another barrack. When I went there to collect it, it proved to no longer be there. Never mind, it can be sent to me later.

[68] A sheet of paper folded into a triangle, stamped and addressed.

Our daughter's illnesses distress me. In the last letter (the 12th of December) Marina writes that Elza had abscesses over her entire body; and you told me on the telephone about her scrofula. My poor daughter! She eats well, judging by the letters. The problem of firewood continues to worry me. I want to try to phone you on Tuesday the 29th of December at 10.00 p.m. True, we agreed to call on the evening of Wednesday the 6th of January at 10.00 p.m. However, first it is inconvenient for me, since I am working during the day shift that week (from the 3rd to the 10th of January). Second, it seems to be a very large interruption. But if during these days (the 29th or 31st of December, or the 2nd of January) it will be impossible to phone you then let the 6th of January 1943 remain reserved.

I will tell to you a little about my life. I work on the day shift this week (from the 20th to the 27th of December). The most difficult day was Monday the 21st, when a sharp frost of -33°C came, and I was required to spend the whole day in the cold to sort and mark the bricks, and other dull business. In general, these days, the job of quality control inspector is becoming stressful. It is necessary to sample, to mark bricks, to watch over the sorting and to check that the bricks are dispatched correctly. But as, in spite of my efforts, the bricks that are produced are of poor quality, so our small but feisty chief threatens and reproaches me. Luckily for me, from tomorrow I move into the night shift again and hence the following week (from the 27th of December until the 3rd of January) the reproaches and threats will fall onto the head of my partner — the second shift manager in the annealing shop, Pavel Ferdinandovich Kromer (a former assistant professor in the physics department of Sverdlovsk State University).

As for my subsistence, this week I was occupied with purchasing bread and exchanging it for potatoes. I bought three 500 g bread rations for 180 roubles and two 800 g rations for 170 roubles, and exchanged them in several stages for two and a half buckets of potato. However, my reserves of potato are rapidly running out.

Since the 22nd our bread ration has decreased to 550 g a day. Therefore I eat more potato accordingly.

The following week I am considering starting to exchange money–bread–money in order to use the additional value to arrange the exchange money–bread–potato. These concerns about food take up much of my time. I do not worry about January too much, since I have about 500 roubles spare. But, as the saying goes, 'we will look' further. I will sell some more junk. On the whole, I do not think that I will die. How do you like this husband-businessman?

Speaking between us — a mediocre husband. I cannot even help my family. In your place I would cry, give up, and look for another husband. Of course, it's only a joke, Lidusya. But it could be possible to draw this conclusion from your long silence, if I did not know you so well. Kiss you strongly.

Your Armin.

Letters 44–45

Thursday 31st December 1942, 2.00 a.m.

Dear Lidusenka!

The last day of 1942 has begun. 1943 will arrive in 22 hours. It's 2.00 a.m. Our usual four got together in the quality control office: laboratory assistant Rikert, quality control inspector Fridrikhsen, quality control inspector Stromberg, temperature master Lir.

We will be on the night shift again during the New Year night, and are going to celebrate the New Year by collectively eating one entire bucket of potatoes and drinking coffee with sugar. They gave out 450 g of sugar per person (in December). I was given the task of acquiring the aforementioned amount of potatoes in exchange for sugar. There will be a New-Year amateur concert on the evening of the 31st in the brick plant's club. I will also take part in this concert as a jazz pianist (!) and accompany one of the violinists during their performance.

I drew a picture for our daughter: Elza with a gift under a Christmas tree. Will the censor really paste over this picture too? Actually this drawing came out less successfully than the previous one (Elza cooking on a small stove), but nevertheless it will be offensive if the censor appropriates this sketch again.

I phoned Modestus yesterday. It's a pity that you could not join the call. I was waiting near the telephone for more than two hours. They connected

The sketch drawn by Armin and mentioned in the letter above depicting Elza beside a Christmas tree.

me with Modestus only during the first hour of darkness. The audibility was good. I am waiting for the photographs of my daughter that Modestus promised to send. Please remind him about this promise. Modestus said that Elza was inoculated (against what?), and her temperature rose because of this, which prevented you from visiting Modestus at the time of the telephone conversation. I do hope that this is not a serious disease?

My mood is sufficiently good. I pin some vague hopes on the New Year, 1943. I want to believe that in 1943 I will say goodbye to the brick plant and return to my native family again. Sasha has the same hopes. I expressed these hopes very keenly in a poem I wrote recently. I will send you a copy.

I still have some money, about 500 roubles, so thus far I do not think about ['the hunger' is crossed out] that time when it will run out. I will try to defer the moment of its shortage as far as possible via the transformation of money–bread–money. So these concerns about self-subsistence and thoughts about your life are the main part of my existence. Any thoughts I may have had about an improving my qualifications, about preparing for a full doctoral thesis are obliterated now. I was a little sad when I received your message that Schur and Vonsovskiy had defended their full doctoral theses, while I spend my time fruitlessly. But it is not entirely correct to say this. If I have been delayed a little with respect to my scientific development, then I have learned much about understanding life and people. I consider your thoughts about our future life to be completely correct.[69] In fact, it is not possible to only live by some 'high scientific ideas', especially as I am only an ordinary person of average ability in scientific matters. Therefore I have decided to consider myself in this way in the future and, first and foremost, to direct my activities towards providing you and our daughter with all that is necessary for a calm and wealthy life. Only after this will I think about my personal scientific progress and about the possibility of developing science of my own. Such reflections come to me frequently these days. Now I have finished my last letter to you in this year. The following — in 1943 already.

Your husband.

We celebrate New Year always
With large convivial association.
Some different snacks were on a table,
And grapevines sparkled on the tablecloth.

Our major hopes were strongly bound
To a young New Year every time.
With what a trembling we awaited
It'll bring some happiness to us.

We cheered ourselves as in a childhood,
Enthusiastic speeches were said...
It is pleasant to recall those dreams,
And meetings where friends were glad!

---

[69]Father explained to me that Mum often rebuked him for his unpractical ambitions and so forth — EAZ.

Now war is blazing everywhere.
And our gifted nation runs
The combat of unprecedented scale
In victorious nineteen forty-three.

But there is no grape vine in here,
And there is no cake at all,
It does not smell of tasty pies,
No cheese or sausages of course.

Salami does not excite my eyes,
And does not attract with a fatty look,
But only empty plates and bowls
Sedately stand and wait for food.

We cooked a bucket of potatoes,
And each brought some coffee along,
The bread rations brought together —
And threw a party all our own!

Though our meal is a little scant,
Though pure water is a drink,
I drink to our homecoming
In sweet and loving families!

This thought excites me to trembling,
And I am too drunk of thinking so.
My hand briskly raises
A clay cup filled with water to the top.

I drink to widespread victory
Against the retreating enemy,
For year later at a dinner
The smell of pie again to see.

There will be glasses and decanters
On a white and perfect tablecloth.
And different vines in those glasses
Will play a rainbow colour tones.

Sausages, loaves, and even butter,
Biscuits, some cheese and chocolate,
Jam, cakes, the foreign pineapples
Will show the real life for us.

We will forget all our torments,
Misfortune is hardening our mood.
Friends, let us shake hands of each other,
And go to happiness and ... food!

Alexander Lir,[70] night of the 30th to 31st of December, 1942, Nizhny Tagil.

Letter 46[71]

Sunday 17th January 1943

My darling!

Recently Rikert's wife arrived here. She leaves the day after tomorrow, so that I am sending this letter with her, and she will send it to Sverdlovsk. I would like you to meet her. In my opinion, she's a nice woman. On the other hand, you could both stay in touch in case the chance of a trip to Nizhni Tagil arises. Just in case, I will write her address down for the second time. Nadejda Leontievna Pavlova, Siberian circuit (continuation of Decembrists' Street), the eighth kilometre, woodworking complex, house No. 8, opposite the Agafurovsky dacha. It is a 25–30 minute walk from the last tram stop in the University Area of Sverdlovsk through the forest towards the woodworking complex to Agafurovsky dacha.

The address of Rikert's wife's sister: Alexandra Leontevna Bessonova… […]

Has Lisa Chernobrovkina been to see you yet? I sent you the shoes with her. She promised to visit you. I saw Lisa on Monday the 11th of January in the morning, and by perfect coincidence it was the same day that she wanted to go to Sverdlovsk.

---

[70] Translation from Russian by ASK.

[71] Written on 19 pages of one-quarter sheets.

nervous and physical energy. I have hardly any free time left. I am entertained only with music — playing the piano in the string orchestra — since there are concerts in the brick plant's club planned for the 11th of April and the 1st of May. Yet I get so very tired that even music brings me little cheer.

I now chase two basic desires: to eat and to sleep. It is even more desirable to sleep than to eat. Concerning food, the following regime is now established. I take skilly without any bread in the morning at 8.00 a.m. During the day at 11.00 a.m. (dinner) there is another skilly without bread. At about 2.00 p.m. I eat half of my bread ration (275 g) and a 'jelly' (see further). A third skilly is taken in the evening at 6.00 p.m. (supper). Around 8.00 p.m. we four get together in our room. The person on duty heats a plate on the coals, and we 'dry' (i.e. we eat) bread — I eat my remaining half a ration (275 g) with salt. We cook 'jelly'. This dish is prepared as follows. Usually during the day we are able to obtain an additional pot of skilly. I cut up two or three potatoes into this skilly including the skin (in order to preserve all of this precious product); and after the potato has boiled I add a quarterof an onion and some flour with water (one and a half table spoons of flour in half a glass of water). Then the contents of the saucepan are poured into a clay plate and this is placed under a bed until the next day. During this time the contents of the plate thickens into the form of a jelly. At present I also receive the so-called 'one intensive'.[93] This means I receive an additional amount of a little of the thicker soup in the morning (four to five tablespoons of soup made of fish or scraps of meat). Sometimes a tablespoon of salt or some chopped mushrooms are added to the evening skilly. The composition of the skilly changes. The basic components are cabbage, croup, beetroot, potato, occasionally fish or meat scraps. But the main component is water.

I sleep at odd moments. I must accept loads of bricks at 1.00 a.m., 7.00 a.m., and 4.00 p.m. Furthermore, the loading of the brick onto the railroad cars often happens at night. It is essential to carefully ensure that the requested amounts are correctly loaded. That's how I live. No time remains for anything besides work, sleep and food.

Kisses to you and daughter.

Armin

---

[93] A concentrated, additional portion of food — EAZ.

Letter 56[94]

Wednesday 21st April 1943

Lidusenka!

Yesterday your letter (dated the 5th of April) was delivered. Thank you. I did not expect that you would be able to write a letter so soon (!) (your last one was written on the 13th of March). This is ironic. I understand that you have many troubles right now. Elza returns home earlier from the kindergarten (at 5.00 p.m.), and you have to feed her. It is also necessary to obtain some seed potatoes for planting in the vegetable garden. Can the UBAS provide you with 5 kg of potatoes? Did they give you a piece of land to grow vegetables in UBAS or the UII?

I am very interested in your house-moving matters. I hope they are solved to your satisfaction! In any case, there is steam heating supplied in the Vtuzgorodok district,[95] and it is close to where you work. You would always be able to come home in the middle of the day to eat and to call on Elza. Have they given an apartment to Mum yet?

It is also good to learn that they have given you one ticket to use in the dining room throughout April (without any reduction in your food coupons, I hope?). It greatly distresses me that Elza is thin and that she is often ill. Maybe she is not eating well in kindergarten? I received Elza's drawings (dated the 11th of April) very quickly (on the 16th).[96] Now you can send more of our daughter's artwork.

---

[94]Letter number eight of that year in AGS' numeration.

[95]The campus of UPI, where Lydia and Armin lived that time.

[96]My father saved this picture and put it in an album for me. It is drawn on a piece of paper folded up into a triangle, it is supplied with a stamp and is sent to:

N. Tagil. P.Office 2, Post Box 67. To Stromberg Armin Genrikhovich.

Text, by the hand of Elza: 'ELZA STROMBERG FATHER MAMA MAMA'

Text by the hand of LMS: 'Elza sketched it sitting in the dining room. Now let us send it. Write to us, dear Daddy!'

Text by the hand of AGS: 'It was sent on Sunday the 11th of April. It was received on the 16th of April 1943.' — EAZ.

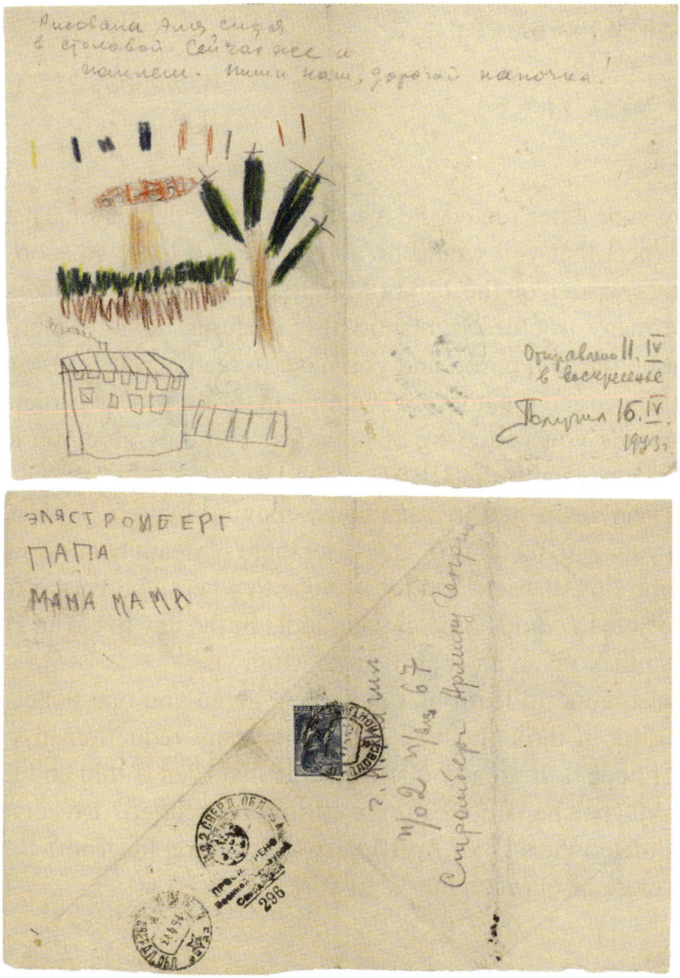

Elza's drawings to her father mentioned in the text above and in footnote 96.

I have also just received one further letter from you (written on the 6th of April)! My happiness and surprise know no bounds. You received the reference. You did not have time to use your food coupons to obtain goods (1.6 kg of meat, 1.2 kg of cereals). It is good that they marked you down on the list to receive an additional ration of potato. If Elza could stay at the kindergarten until 8.00 p.m., this would very much lighten the load for you.

Several words about the sale of my things: after long deliberation I sold my black woollen pants for 5 kg of bread which equals ten rations of

500 g. At first I wanted to put off selling these until the autumn in order to barter them for potato, but hunger can make one do anything. It is now the second day that I have been able to eat 1050 g of bread per day. I'm in perfect bliss. I eat 500 g during the day between dinner and supper, and another 550 g after supper. And that's nothing to me! ... So until the 1st of May my buyer will give me an additional 500 g of bread, daily. I decided that it makes sense not to try to extend the lifetime of this bread (for example, not to dry rusks), but to eat the extra 5 kg of bread over ten days, and in doing so to immediately avoid any potential barrier to transfer myself from the state of a weak person to the state of a normal, healthy person. This seems to have had positive results, as the swelling on my face has begun to fade away and the weakness in my feet has begun to disappear.

Concerning the summer suit, I wanted to ask you to send it to me, and also the pair of white jackets with the separate collars, the pair of neckties... The fact is that my 'dark-blue suit' has finally become worn out, and I look like a ragamuffin. However, most of our engineering staff has decent clothing, and this is important to maintain one's 'official career' under our present circumstances. So Lidusya, please send me the clothes mentioned above at the earliest opportunity. In autumn I can exchange this suit for some potatoes in case of dire need. I think it will be easier to do this in autumn.

Over these past few days I have been very engaged with music. Each and every day our group of 'jazz players' gets together in the club to rehearse our program. Today (the 21st of April) at 10.30 p.m. dress rehearsals (!) will begin, and on the 24th of April we will perform on tour (!) at the *Tagilstroy* club. Of course, we will also perform on the 1st of May. They promise to give out some clothing — of the jazz form: boots, trousers, a service jacket. It might be that some food comes our way. Thus, it seems I can have both a pleasant and a useful pastime in music.

I'll end here. It's 8.00 p.m. now. I am sitting in the quality control office and writing this letter after the end of the day shift. I decided not to go home after supper, since nowadays it's difficult to leave this zone in the evening, and the jazz rehearsal must start at 10.30 p.m. in the club. At 9.00 p.m. I plan to go to the cinema to watch a documentary film on the defence of Stalingrad and Leningrad.

Yesterday Carl Fridrikhsen returned to our room after one and a half months in the hospital (maybe you remember that he helped me to buy those felt boots when you came to me in August). His diagnosis is malaria and gastric catarrh. He is terribly emaciated. Fortunately, he received a parcel from home. This will help him to pick up. If only I do not fall ill, damn it!

Lidusenka, please write more frequently. It is so pleasant to receive your letters, and makes me the envy of my other comrades.

Kiss you strongly.

Armin

Dear daughter Elza!

I received your pictures: a mushroom, a tree, a house, Dad and Mum wheeling Elza in the carriage. Please draw something else for me and ask Mum to send it to me. How do you spend your time at the kindergarten? How do play you with Leo and Misha?

What books are you reading? What new toys have you got? This is all very interesting to me. Strongly kiss you, my dear daughter.

Your dad.

Letter 58

[Letter 57 is missing.]

Wednesday 5th May 1943

My dear wife!

So, today is the birthday of our daughter, Elza. Exactly five years ago (on the 5th of May 1938) I sat at 3.00 p.m. in the maternity ward on Bolshakova Street and awaited the results with great agitation. They told me that you were in the operating room. Finally, the doctor left and he said: ‘Everything went smoothly, your daughter was born; a pretty, strong child weighing 3800 g.’ Two days later I saw my ‘pretty’ daughter through the window in a cot, with her chamfered skull, and eczema on her face. The May sun lighted upon her. These days are so memorable to me!

On the 2nd of May 1938 we could still walk as a family to the culture and leisure park. We were so carefree! You stayed for about 20 days in the hospital. How often I approached your window, climbed up to the cornice and spoke with you. The black cherry flowered under the window. Yesterday the snow melted, and I recalled that five years ago in 1938 I went to the UBAS in the morning wearing only one white jacket, and suddenly the temperature began to drop sharply during the day. Shivering with cold, I ran from the tram to the hospital with a parcel, since visiting hours were only from 3.00 to 5.00 p.m. From there you arrived home in a car. Our first photograph of our daughter was taken at the age of 19 days (!). I can clearly see our daughter's album. It's a pity that I cannot continue to add to it. Lidusya, please put several of our daughter's drawings in it, and the photo taken by Modestus. It will be so pleasant to recall about Elza at this age.

The last letters from you were delivered on the 20th and 22nd of April (written on the 5th and 6th). On the 27th of April we spoke on the telephone (for the first time this year!). Today I visited Lisa's office. They said that she is on a mission in Sverdlovsk till the 10th of May. Probably she will visit you, and you will be able to send me money, fats, jackets and a suit. On the 2nd of May I dedicated a substantial part of the day to cannibalizing my black trousers to use them to repair my dark-blue trousers. On the advice of Carl Germanovich Fridrikhsen I decided to return them to the tailor's repair workshop to put black rhombic-shaped patches on the elbows and to sew on some black patches of the correct shape onto the backs of them. Also I will ask them to hem the lower part of the trousers. I'll walk throughout the summer in them somehow. I completely 'ate' my woollen black trousers. Today was the last time I ate an additional 250 g of bread: 250 g in the daytime and usual ration of 550 g in the evening. Tomorrow I am going to purchase an additional 500 g of bread (for 50 roubles). Then I will be satisfied (800 g of bread) on the 7th and 8th of May. Sadly, I suppose I'll have to live once more on just 550 g of bread daily. The nourishment in the dining room leaves much to be desired. The menu until today has been: breakfast — liquid skilly with peas + two to three tablespoons of porridge made of ground wheat with 'hooves and horns'; dinner — liquid skilly with more ground wheat + a liquid pea porridge (two table spoons). Supper — skilly with some traces of pea at the bottom… Bread is our main fuel. If you could send me 300 roubles, then I could purchase 3 kg of bread (0.25×12), i.e. over 12 days I'll eat 800 g ($550 + 250$). I'm so tired of this arithmetic!

I have a lot of work in the past two days. They gave me 30 workers to re-sort and select pipe-brick.[97] I should give out the tasks and indicate the area in which to store the bricks, monitor the quality of the sorting and finally sign-off the work. At the same time I must do my current work as well. At the end of the day I can barely drag myself to my room and fall into bed until 12.00 a.m. By now I have recovered my breath a little and have sat down to write you this letter. I am very interested in our domestic matters: moving to the new apartment; the maintenance of the vegetable garden and planting of potatoes. How much time does our daughter spend at kindergarten? Did you get a passage into the dining room of UBAS. Did you obtain potatoes for the families of those who have been mobilized?

Write to me some more.

Your Armin.

Letter 59

Friday 14th May 1943

My dear Lidusya!

Forgive me for the long silence. I put aside this letter until I received a package from you. At the beginning of May I called Lisa at work. They told me that she was on a mission in Sverdlovsk until the 10th of May. On the 11th I visited Lisa at work and received your message. Thank you darling. You sent everything that I asked you for in the note: butter, fat, tinned food, a loaf of bread, my white trousers, the striped jacket with the collars and the necktie.

Special thanks for the butter. Pass my regards to Serge Vonsovsky and my appreciation for delivering the parcel.

Our diet has improved considerably during May and the daily standards have increased. I receive 650g of bread. They nearly doubled the ration of products in the dining room. In addition to the first course they also give us a little porridge during the evening supper. Our engineers who work in the production unit even receive 800–1000 g of bread depending on the percentage of production. Workers on hard tasks receive a bonus dish and so forth. Everybody has begun to cheer up. Stakhanovites — 300-percenters — have

[97]The bricks used to build shafts had to be of a higher quality — EAZ.

appeared among us. It has begun to be a lot more cheerful at work. Lately I have fairly often succeeded in obtaining even more additional skilly, on at least three occasions (by pulling strings), since sometimes friendly engineers are on duty in the kitchen quite often. Yesterday, for example, Carl was on duty from our room, and of course he did not stint in giving me food. I frequently have an extra dinner at night when I work on the night shift and so on…

I can say without any particular boasting that I'm quite successful in hunting for additional skilly; and I possess great skill in this matter. At some point, in the future, it may be funny to recall these boyish *sorties* of scientists. However, at present we discuss these questions of finding food in absolute seriousness.

Along with matters concerning our diet, I also give much time to the problem of summer clothing (I did not hold out much hope that you would be able to send me the trousers and jacket so quickly). After selling the black woollen trousers in the middle of April I was left with only one pair of padded trousers. I wore out two pairs of summer trousers (dark blue and black) this past summer, so it was improper to wear these even in the territory of our plant and barracks. My dark-blue coat also turned to rags. The problem with my trousers has already been solved. Our dressmaker, Claudia Timofeevna, made me one pair of trousers from two old ones (she did this for me out of turn, taking into account my miserable situation with the padded trousers). She sewed large strips from the black trousers to the dark-blue trousers with her sewing machine: large rhombic-shaped patches to the elbows, and the entire tail end. They look a bit out of the ordinary, but they are completely suitable for working in. The problem of a coat will probably be solved soon as well. Claudia Timofeevna promised (at some point) to make me a coat. During our conversation it transpired that her electric iron had burned out. I repaired it by placing aluminium rivets on the burnt places with a nichrome spiral, included resistors in series so that there would be no overheating. She was very grateful and promised to sew the coat without me having to wait in turn (!) and to line this coat also. So in the next week or so I will show off in my new coat, my white trousers and in a new jacket with a necktie.

Our workload has now begun to ease off after a month of hard work. A special person — a controller-foreman — will now carry out that part of my job (the inspection and loading of the finished products), which had until now occupied the greatest part of my time. I merely remain as the quality control inspector in the kilning shop. Today the controller-foreman, Rotermel, will be on the night shift together with me for the first time (he is a former second-year student of the UII, a metallurgist).

Lidusya, please write to me how your life flows. It is a busy time for you: the end of the academic year, with examinations, checking course work, consultations, digging the vegetable garden and planting potatoes, caring for our daughter and worrying about money. Write to me (at least briefly, while you are sitting in the dining room) how you mange to cope with all these problems. How is your health (your headaches) and the health of our daughter?[98] Please send me the photos printed by Leleshka.

Strongly kiss you my darling.

Your most fashionable husband.

---

[98]Comments on Elza's health and well-being from Lydia's letters to AGS in May 1943, which were preserved in the album of Elza, recorded by the hand of AGS — EAZ:

Elza makes many sketches for you, but I can not send all the sheets. We cooked breakfast with her this morning. The last salt cabbage was cooked with some potato and stuffed with an egg, which Lydia Stepanovna Shubnikova [Armin's cousin] presented to Elza on her birthday. Almost everyone remarks that Elza is very pale, thin and changed. I try to give her the best I have, but she has a poor appetite. Yesterday she took part in a morning performance. She recited a poem and danced together with the others... She is desperate for your return so that you could make a small jug for her doll. She is very fascinated with dolls. Today she ate chocolate that I presented to her on her birthday on behalf of you, and she said: 'When I bite off a piece, it means I am kissing my daddy...' In general, she thinks about you every day. [...]

Yesterday I planted a bed of carrots. We worked late into the evening together with our daughter. She helped me to eject some stones from our vegetable garden. Today our daughter's ear was sore the entire night. Now I run to dig the vegetable garden without the chance to sleep myself, and must go to work in the evening. There will be no vacations in the summer, so I will not visit you. [...]

Elza has become very adult and clever. She loves to receive letters from you. You write so little about her, and she always asks that I read your letters aloud. [...]

We sit now with Elza in the dining room and have dinner. Elza got into a sweat and undressed down to her shirt. We walked from Vtuzgorodk. I am tired. The vegetable garden is finished (two hundred square metres). [...]

[Written in the margins:]

How has the housing problem been resolved?

Letter 60

Tuesday 18th May 1943

Dear Lidusya!

Today I have great happiness. Simultaneously, three letters were received from home: two from you (dated the 8th and 9th of May) and one from Mum (a postcard dated the 9th of May).

I again report to you (in case you did not obtain my previous letter) that I received your parcel from Lisa, which was brought by Serge on the 11th of May, as she had visited her husband on the 5th–9th of May. I have already drunk the Hematogen (many thanks to Tosha for that), and ate half of the butter. I eat it with bread with huge enjoyment (a result of the lack of fats in my body). I have not opened the tins of food yet. Of course, the loaf of bread was mouldy and stale (you'd better send rusks, Lidusenka). I have not worn the white trousers and the striped jacket yet because of the cold weather. It only began to warm up again today. Lidusya, do not buy me trousers since now I have three pairs: one white, one in two colours and a green pair. My seamstress in our tailor's workshop sewed the bi-coloured trousers for me from two torn summer trousers: a dark-blue pair (that used to be a suit) and a black pair. Some green camouflage trousers and boots were issued to me recently in the club as a uniform for the jazz orchestra. For certain, I will not only wear these trousers and boots at the performance (!). I returned the dark-blue coat to the workshop to the same needlewoman. Yesterday I helped her to cut up the coat, and today she has already started to sew. After this coat I want to also give her my grey double-breasted jacket to be turned over. Thus, I am provided with clothing for the summer. Tomorrow Boris Viktorovich will go to the city. He will carry two packages to Lisa. At the end of May she wants to go to her husband in Degtyarka. If she agrees (I have not made any preliminary arrangements with her, since your letter was only delivered recently and I have only just learned that my comrade is going into the city), then these bags will soon be with you. Lidusya, if you still have some spare money (for example, the 300 roubles that you wanted to spend buying me trousers) then please send it to me via Lisa. I want to purchase bread before any fresh potato appears on the market.

The food in May was somewhat better. As a time-rate worker I receive 650 g of bread a day, others receive 600 g. Those who make 100–130% of the standard obtain 800–1000 g of bread. The masters working in production (who write orders for the workers) obtain 700–800 g depending on the execution of their plan over the last three days. Lately, I think about moving to work as the shift-master of the kilning furnace for the summer group of workers. But I am also not sure about this, since first, there is little chance that they will let me leave the quality control office; second, it is a less durable position. In October the summer group will be disbanded for the winter again. It is easy to be denounced for leaving work, and given guardroom duties etc. in case of any oversights. Still, I will think and talk about this with the authorities. Nevertheless 650 g and 800 g is a big difference. The meals have also become considerably better. Aside from skilly, they also give us a small portion of porridge and fish at dinner and supper.

I have finished this letter on the morning of the 19th of May. Now Boris Viktorovich will go to the city, and I will hurry after him to send this letter. I will write a continuation of this letter soon.

Your Armin.

Letter 61

Thursday 27th May 1943

Lidusya!

Your last letters dated the 8th and 9th of May were received on the 18th. At that time you had not started work on the vegetable garden. By now you have probably finished planting potatoes. How many hundreds did you plant? How many buckets of potato did you plant? It is a shame that I cannot help you at this busy time, and I must be inactive while you are working hard.

After receiving your message (the 11th of May) I started eating the butter, and at present it has already run out.

Over the past five days I have begun to give serious consideration regarding the matter of how to improve my diet with a given ration. There is only one possibility now: to move from being a quality controller in the kilning shop to the post of shift-master. Then I will obtain 800 g of bread instead of 650 g. Of course, the work is less interesting. But an extra 150 g of bread a day is a serious argument to me. I would agree to the general foreman of kilning's proposal without a moment's hesitation. However, the chief of quality control — Vulakh — has opposed this move most categorically. Now she has begun to think of ways to improve the diet of the quality control inspectors, since the three remaining quality control inspectors may also move to another post that can earn 800–1000 g of bread. Yesterday evening we saw the first tangible result of this matter. They gave us four coupons for supper in the civilian's dining room (in addition to supper in our dining room): porridge with soft fish roe. Of course this is not a substitute (for example in calorific value) for 150 g of bread; nevertheless it is better than nothing.

On the 25th of May I was on duty in the kitchen in our dining room (as the person on duty from the plant). I followed the correct procedure to obtain products for dinner and supper from the store, and checked the distribution of foodstuffs for breakfast. Whilst the food was being served I explained and resolved any kinds of misunderstandings that arose. However, unofficially I tried to look after myself and to feed up my friends as far as possible. My friends left contented. As far as possible I made sure that they received several additional portions of skilly, porridge, fish, etc. But I myself was left very dissatisfied, since I ate a lot of food rather like a small child in the first 8–10 hours of duty, and then during the remaining 18–20 hours I could hardly eat because of strong pains in my stomach, and a total loss of appetite. Well, it was upsetting! Perhaps, this was the first time in a year when I have had no appetite at the precise moment when I could gorge as much as I would like: porridge with butter, honey cake, boiled fish, etc.

It is ridiculous and it is sad. I have not fully recovered, even by today. Soon Carl Germanovich is on duty in the kitchen (another quality control inspector). Surely he will feed me up (this will be in the first days of June already).

That's how I live, Lidusya. I attempted to send you a bag for the rusks via Lisa with a comrade who was walking to the city (on the 18th–19th of May), but Lisa refused to take it, as she will be going at the end of May to visit her husband along with her child (maybe she has already left). Please send me a package of rusks if possible, and also around 300 roubles if you have any left. Do not buy me trousers, since they gave me a pair of green trousers.

[Written in the margins:]

Write about yourself, about our relatives. Where is Volodya? What does [Lyusya] write?

Letter 62

Thursday 10th June 1943

Dear Lidusya!

We have arranged a 'feast' in our room tomorrow on the occasion of Voldemar Ris' send-off. Yesterday a telephone message was received at the brick plant, ordering him to be demobilized from the construction crew as of the 11th of May. The commissariat busied themselves on his behalf, and they sent him back to Sverdlovsk to his previous place of work (he is a designer of turbines and fans). I have sent you a bag for rusks and some music for Marina as a birthday present. He will phone you, and you can visit him. He will tell you about our life and existence lately. Two more people were demobilized from the force several days earlier. Getling is Russian on his documents. Tsvingman is a well-known specialist in wooden construction. These events give me a certain hope for a favourable outcome of the efforts of Mum and the UBAS [to have AGS released — GGW]. However, it must be noted that the commissar who pleaded on behalf of Ris is his former student, so that can be a most valuable circumstance.

On the 7th of June I received your 300 roubles from Maria Ignatevna. I have already purchased a ration of bread (600 g) for 60 roubles. Every three to four days I will purchase a ration of bread in order not to starve and to supplement the standard food's nutritional value. I recently attempted to move to the post of shift-master of the kilning shop. The chief engineer promised to support me in this. But the chief of the quality control shop (Vulakh) downright refused this. Nevertheless I made a statement. The director took the side of Vulakh. As a result the deal ended in disgrace. The director threatened to send me off to the guardroom (!) and to send me off to the coal-face (!) if I would not stop speaking about a transfer. Thus, the 800 g ration has 'smiled' at me (and said goodbye) so far, and I am left with my previous 650 g ration of bread. They excluded me from the list of people on duty in the kitchen on some random pretext (I should have been on duty this very night). Of course, it's very sad. This week I work on the night shift. I obtain an extra *balanda* each night (by pulling strings), and a usual daytime meal. That's how I live, Lidusya.

I have prepared a musical gift for Marina in my free time. Five pieces of jazz were transposed for the piano. Three of them are with words: 'In a Cottage', 'Parcel', 'Farewell' and two additional pieces 'Oxford Circus' and 'Waltz'. I am now reading Goncharov's *Oblomov*.

I end here. It is time to go to work. Write about yourself.

Armin

[On a quarter sheet of paper:]

Lidusenka, please write to me in detail about our daughter. Will she go to the dacha this summer? Did you plant potatoes, carrots, etc.? Do not rely on my speedy arrival. I do so want to arrive in order to help you overcome the everyday questions: apartment, firewood and money. This is so foolish. You have not one single minute of leisure, and I read fiction in here, I am occupied with music… I hope my return to Sverdlovsk will become possible at least before the winter. Kiss you. Your husband.

My dear Elza! My dear daughter!
Are you healthy?
Do you obey your Mum?
Will you go to the dacha in the summer?
Draw pictures for Dad: tanks, aircraft, guns, steamships.
Ask Mum to send me your drawings.

Strongly kiss you, my dear. I will return to you as soon as the war ends.

Your dad.

Letter 63

Saturday 19th June 1943

How do you do, my dear Lidusya!

(Carl begins all his letters with these words; he writes to his wife Alya almost every day, and she writes with the same frequency.)

My life and work in the quality control office flow quietly again after the troubles connected with my recent attempt to move from being a quality controller to the kilning shop. The change in the plant's management — a new director and a new chief engineer, the fourth change so far — have not had any influence on the quality control service. Today is the first warm day, and it is possible to walk without a coat.

I re-read your letters and I will answer some questions.

Sasha Lir is healthy. He tells me that on the 3rd of May he sent a letter to Vera asking her not to write to him (!).

I saw driver Schmidt recently. He works in our garage. He says that has not received a single letter from his wife this year (!). He worries about how Dusya is surviving.

I received your 300 roubles and have already bought four 600 g rations of bread for 240 roubles.

Several questions to you: Have you seen Ris? Please find out about any troubles relating his return to work. How is my 'deal' being pushed along?[99]

Will Elza go to the dacha in the summer with her kindergarten?

Has the potato frozen or not?

Marina wrote that there was talk of a purge of 'outside'[100] children from the kindergarten. I hope that Elza will manage to stay.

Have you received any letters from Krasnoufimsk from your mum and Lucy? The same question regarding letters from Moscow from Granny Olga, from the Shubnikov family, from Elga, from

[99]The first explicit mention of Armin and his mother's attempts to negotiate his release and de-mobilization — GGW.

[100]The jargon of Bolshevik party members referring to those children whose parents were not party members — EAZ.

Volosik, from Vladimir and Peter? Where are Golubovskaya and Reyson?[101] How is Tosha's health, Michael, Leo? Please write about Elza in detail.

What will you do with the summer? Will you be able to visit Nizhni Tagil?

My life: I am in a state of indifference to any external goings-on. I try to entertain myself by reading literature, with music, and by reading letters. I try not to think about food. I read selected novels of Chekhov. At the moment I am writing some music for 'The Dark-blue Headscarf' and 'The Seagull' songs, but with new words. I saw the *Valery Chkalov* film. I sleep now on a real bed (!). After Ris left they took the two-storey bunk out of our room. The absence of outside news has a depressing influence on one's mood. It seems that I'll sit in this Nizhni Tagil for all eternity.

Please write frequently.

Your husband.

Letter 64[102]

Sunday 4th July 1943

My dear wife!

Your last letter was dated the 29th of May (!). From the last letters of Marina (the 18th and 22nd of June) I already know that in June you prepared to send Elza to the dacha, and without doubt were working hard at the Institute. I can justify your long silence by these facts. You did send Elza to the dacha on the 25th of June. Your conversation with Ris regarding the circumstances of his release interests me greatly; what led to the successful outcome of his returning to his work? He left for Sverdlovsk on the night of the 14th to 15th of June. Is it true that he did not phone you?

---

[101] Artists who evacuated to Sverdlovsk and lived in the home of the Stromberg family — EAZ.

[102] A paper triangle with the address and a stamp worth 30 kopecks. It was examined by military censorship.

I gave him the telephone numbers of Modestus and Shlygin. How did he get on with his registration? How is the garden coming along? What do you plan to do at the Institute in July and August? Will you be able to visit in August for several days so that I can see you? Indeed it will soon be a year (!) since we last met (this was on the 17th to the 21st of August 1942).

Much water has flowed under the bridge since then. Especially taking into account the rainy weather this summer (!). Nevertheless this year seems to me to have flown by rapidly. There have been no significant changes in my life. My mind has not been occupied with higher thoughts, and was merely inventing different ways in order to obtain potato and bread to hold body and soul together. In order not to let the brain turn to mould I finally became occupied with music in June: arranging jazz pieces for the fortepiano (as I have mentioned to you already, in May the leader of jazz band left the confines of the brick plant and the jazz band was disbanded). Second, I'm occupied with reading literature that a roommate obtains (having a pass to go to the city) from the library of the machine-building institute. I read a volume of the selected compositions of Saltykov-Shchedrin and Chekhov. I now read Gor'kiy. In recent days I shook off the dust from Guggenheim's *Thermodynamics*; now I am nearing the end of its study with a fresh flow of enthusiasm.

Lidusenka! If the opportunity arises then please send me the following books to Lisa's address (57 Newspaper Street): 1) *Thermodynamics* by Lewis and Randall; 2) Pletnev and Sklyarenko — *A Book of Problems in Physical Chemistry*; 3) The same by Knox. I will work on my thermodynamics in these coming years in order to teach it at an institute after the war (!).

I have made friends with a worker in the club by giving music lessons, and so I now have a few strings to pull to get free visits to the cinema. Hence, two to three times during the week I go to the cinema in our club. Recently I watched a film, *Mashenka*, and was very happy.

Lidusenka! Write to me some more about Elza. I brought the album (notebook) for her, where I put extracts from these letters about Elza. I have to note that most of material was taken from the letters of Marina. Lidusya, please write to me about the different ridiculous phrases, tricks and questions of Elza.

What do your relatives from Krasnoufimsk write about? I wrote a letter to Tosha.

Kisses, Armin.

Letter 66

[Letter 65 is missing.]

Wednesday 14th July 1943

Dear Lidusenka!

The last letter from you is dated the 29th of May! How do you explain this disgrace?! What is the reason for such a long silence? Did you forget about the fact that you have a husband who is interested and worries about the lives and health of those he loves? Fortunately, I have received a letter from Marina in June (the 22nd), so I know that you and our daughter are healthy, and that Elza went to the dacha on the 25th of June with the kindergarten; and that you had to rush to prepare your daughter's clothes.

What are you doing at the Institute now? The classes are surely finished by now. The students must have departed to work on the farms. Did they send you to the kolkhoz? Do you do any scientific research? Is it possible for you to come to see me during July/August for several days? It has been almost a year since we last met, but it is only a six to seven hour journey on the local train (!). You are now free from looking after our daughter.

Darling, please think about this and write me your thoughts on the matter. We have much to discuss about the present, about the near future and about the more distant future. However, we can hardly say everything in letters and on the telephone. Next week I will work on the night shift, so I will attempt to phone you. It would be so nice if you came to visit! We could rent a comfortable isolated room… Have you forgotten last year's woodshed? But indeed winter will begin again in three to four months, the third winter of war!

I am writing to you in the corridor of our barrack, whilst guarding some belongings stored in the corridor. This is because they are redecorating our room: plastering and whitewashing. I work on the day shift, but yesterday I was given two days off due to a 'mutilation of my left foot' [...]

Letter 67

Friday 16th July 1943

Lidusenka, darling!

The same letter dated the 29th of May is still in front of me; I have not yet received a single letter 'from home' for so long. Marina wrote the last letter 'from home' around the 22nd of June, i.e. almost a month ago. Yesterday Carl was in the city and purchased two single telephone coupons for me. Next week I will accomplish a telephone conversation with you. I will learn what is happening at home, and I hope that it will be possible to persuade (!) you to visit this unloved and forgotten husband.

We now have an infirmary in our room. During the repair of the room Carl was painting the windows and got a huge splinter in his finger. Therefore he walked yesterday to the city hospital for an operation. They gave him leave from his work. Today he makes up for lost sleep, and tomorrow he plans to go looking for mushrooms. Alas! I have to wait because of my swollen foot. I also have leave from work, and generally I'm very contented. I feel like I was at a health resort. My foot does not ache in a sedentary or recumbent position. It is only painful to walk on. Apparently, the damage is minor; it is just a simple bruise. All the toes on the foot move, but the swelling is enormous. First the swelling was near a toe on the place where the trolley crushed my foot, and now it has spread to the heel, and the entire foot is swollen, so that even the bones cannot be discerned. Of course, my boots no longer fit, and I hobble in my rubber boots. I walk to our dispensary every other day. The woman is a medic; she's a young person with a magnificent shock of hair (dense on the head, and empty in the head!). She smears my foot with iodine and recommends taking warm baths. That's why I go in the bath (!) three times a day, and at the same time as I warm my foot I also wash from top to bottom. Three times a day! Lidusenka, you should remember how you could hardly force me to have a bath in Sverdlovsk once per month! But there were queues ... here the bath is quite close by, and it is empty at certain times of day (breakfast, dinner and supper times, mainly). The best thing is that hot water is always available and free of charge. My comrades bring me food from the dining room to my room. The weather is excellent. Yesterday they finally cleaned the room after the redecoration. Therefore, today is the first day that I can completely relax. The whole day I read... Guess what? A book by the academician A.I. Oparin, *The Origin of Life on Earth.* Perceptibly an interesting book! I took it to read from Paul Emilevich Rikert. I never thought that I would read a book on this comparatively distant branch of knowledge, as biochemistry is to me such an interest. Of course, primarily this is the merit of the author, since the book is written for a sufficiently popular audience and it mixes entertainment together with the serious contents of the book. You can have a talk with Tosha and Nina[103] about this book. If they have not yet read this book then advise them to read it from me. Also I recommend that you read it too, although you do not love chemistry...

Lidusya, how do you like the paper that I write this letter on? This paper was taken from the cement bags. These cement bags are specially brought to the kilning shop (because of the absence of any other paper) to connect separate pieces onto the paper panels that cover the spy-holes in the Hoffmann kiln. I do like this paper. It is bad that it is so crumpled. But after smoothing it out with an iron it becomes entirely smooth (this sheet was not ironed).

That's how I live, Lidusya. Our nourishment is comparatively good. True, they have reduced the bread ration in the past ten days again (temporarily?), so I receive 550 g. But the skilly is better in May than it was in winter and in spring. The food is not too diverse. The main products are ground wheat and herring. In the morning they give plenty of liquid skilly made of ground wheat. The same skilly is for dinner, plus a little porridge from the same oats. More thick skilly is served for supper, plus some herring (70–100 g). When I am on duty at night I eat an evening meal and usually succeed in eating another in the daytime. I have not managed to obtain additional portions of skilly lately, since I do not visit the dining room, and furthermore, unknown engineers are on duty now. However, I do not feel such a strong hunger as I did in April, when I sold my black

[103]Who were both medics — EAZ.

woollen trousers. My sedentary way of life also contributes to this, which started after the incident with the trolley (since the 6th of July).

It is a pity that you are not here during these days when I am free from work. But I think that when you arrive I will know how to be excused from work one way or another in order to spend those days with you when you come to stay.

I await a letter from you, Lidusenka. Please write to me more frequently. Now there is not so much work as in spring, and even our daughter is away at the dacha.

Lidusenka, please think about visiting Nizhni Tagil.

Armin.

Letter 68

Nizhny Tagil City,
Monday 19th July 1943

My dear wife!

I recently returned from visiting the surgeon at the city hospital. The medic in the construction crew gave me the directions, since she was uncertain whether the joints of my left foot and toes are damaged or not. Everything is in order! The bones are not broken. However, the injury is still bad and the ligaments are damaged a little. The surgeon prescribed me to take some lead ointment at night and warned me that the foot will be sore for one to two weeks more. I'll probably have to request three to four days of additional leave from work, since the foot has begun to badly ache from walking. I do not get bored (!) sitting at home without any work to do. My comrades still bring me food from the dining room. Therefore, I can sit at home for the entire day. I leave this area only to visit the medic in our dispensary, to go to the washhouse to take hot baths for my foot (they told me in the hospital that it is not necessary to do this), and to cheer up and thoroughly warm the foot in the sun in a glade. The weather is excellent so that it has been like having a rest at a health resort in July (!). The only pity is that you are not here now! We could spend several days together.

This evening I will try to book a telephone conversation for the 21st of July at 9.30 p.m. at Modestus' telephone number. It would be good if you could visit in July or in August! I can foresee three difficulties. The first — you must be relieved of your work, business and our daughter for several days. Mum writes that our daughter must return from the dacha on the 1st of August. The question then arises as to where to send our daughter, since I think you would not have time to pack up before the 1st of August. Still, Elza is grown up now, and I think that it is possible to leave her for several days with Mum (probably, her school holiday will still continue through August) or with Tosha.

However, I think it will not be a big hassle to take some time off from your work (to take leave for several days without a salary in order to deliver winter uniforms for your mobilized husband). The second are material considerations. Yet this should not disturb you, Lidusya. Do not bring anything for me. Take only as much as you need for your own subsistence. True, they temporarily reduced the standard of bread lately (I receive 550g of bread daily), but the skilly is better even in comparison with last summer (when you were here). I do not feel hungry any more. Once a week I sell a ration of bread (for 80 roubles) and spend this money on green onions (10–15 roubles for a bundle) and vegetable tops (!) of beet, cabbage and the turnip (35 roubles per bundle), and thus I stock up on vitamins for the winter (!). As for buying potatoes in the autumn, I pin my hopes on my white trousers (I will have to sell these in summer, since autumn is no longer a season for them) and on the striped jacket. But this is a subject for the future (September and October).

The third and the main thing is to get permission for the passage and the purchase of a ticket. In the next few days I will attempt to obtain a permit from the construction crew at the address of the Sverdlovsk police station so that they will permit you a return passage by railroad to the station in Nizhni Tagil in order to bring necessary clothing to your mobilized husband. Recently, one comrade obtained such a permit. I will try as well. Although, they say that the commissioner signs such permits very unwillingly. But if I do not succeed, then I hope that you will succeed in obtaining some mission to visit Nizhni Tagil with the assistance of Mum, for example, a mission to the machine-building technical school. Or it can be made with the assistance of Isaac Constantinovich[104] to any of the Nizhni Tagil plants in connection with your research work. Please think about this, Lidusenka. Indeed if we do not meet in July–August then your later arrival would hardly be possible due to your classes in the Institute, and we will have to wait again for the autumn of the next year. Still, we cannot know how the situation will change. I may have to leave to go to another plant, and then it will no longer be so convenient to communicate. Indeed, now it is only a six to seven hour journey by train, and it is easy to obtain a ticket for the local train. I would very much like to see my daughter too. Many wives arrive with their children. Recently a wife arrived to visit the political instructor of the column from Sverdlovsk with a five-year-old daughter. But I'm afraid that Elza would catch a cold on the way, or she could get any infectious illness (if she were to suddenly fall ill with dysentery then the entire effect of the summer and the dacha will come to nothing). Therefore I voluntarily agree that it is not worthwhile for you to visit me in Nizhni Tagil together with our daughter. You must describe her to me in detail and this will be very pleasant to me as it is.

However, there has been no letter from you for so long, my dear! It has been more than one and a half months (!) since the last letter from you. Have you fallen out of love and forgot about me completely? If you put aside a letter before obtaining a final answer on my 'case', then it is completely pointless. I am already reconciled with the thought of spending one more winter in Nizhni Tagil. It might even be preferable (?). I am interested: did they register Ris in Sverdlovsk, or did he have to leave? You have told me nothing about your conversation with Ris. Did you visit our daughter on the dacha at Shartash? How are the matters concerning the move to the new apartment? How about firewood and potato, supplies of goods and products on the ration cards? What about access to the dining rooms? Have they mobilized you to the harvesting campaign or others job? I hope that the letter from you is on its way to me and I will receive it soon (?). Kiss you,

Your forgotten HUSBAND.

---

[104]Kikoin — head of the physics department — EAZ.

Letter 69

Nizhni Tagil,
Wednesday 4th August 1943

Dear Lidusenka!

Yesterday I received your postcard dated the 27th of July. What is wrong with your liver? Maybe the diagnosis is not correct? Was the doctor experienced in these things? Well, it is good that they gave you a few days off — although you will have little rest.

The day before yesterday (the 2nd of August) I requested a telephone call with you for this evening (the 4th of August at 9.30 p.m.). At 9.30 p.m. I went to the intercity phone station to ask about this booking, and they suddenly told me that there was no such order! This version of an 'unsuccessful' conversation has never happened to me previously. Of course, I was very upset, and immediately re-booked the call for the 6th of August. By the way, in one of the letters I asked you for a phone call at the apartment of Modestus on the 6th of August at precisely 9.30 p.m. However, it remains to be seen whether this conversation will take place. Therefore I decided to write down for you what I was going to tell you today on the phone.

Of course, the main theme of conversation must be the question concerning your visit to Nizhni Tagil. You said in the postcard that this is impossible. But why? Nevertheless, I hope that you are recovering well at the moment. Our daughter is on the dacha. Can you really not be spared for two to three days in August? Indeed it is only a six to seven hour journey on the local train. Many wives live much further away, and even they manage to visit. For example, right now Pavel Kromer's wife, Ella, is here. She travelled from the village of Mityaevo in the Ivdelsky region (to where they were forced to move from Sverdlovsk in December 1941; she taught German in the pedagogical institute in Sverdlovsk). She was travelling for… Five days! For the first two days she travelled by boat along the Lozva River, then 8 km to Ivdel. In Ivdel, at the Board of Public Education, she was able to make up a fictitious mission to the Nizhni Tagil Board of Public Education. Next she rode 60 km from Ivdel to Bogoslovsk on a horse, then to Serov by train, and another train to Nizhni Tagil. And here she is (!!). I have just visited them this evening before attempting this most unsuccessful telephone conversation with you. I am so jealous! Pavel Ferdinandovich Kromer (he was an assistant professor in the physics department of Sverdlovsk State University, who studied with you in Leningrad, one year below you) dreams that his wife could stay and to live in Nizhni Tagil 'for ever'. But this will hardly be possible. Some of the wives have already arrived — for instance, the wife of professor Otto Nikolayevich Bader (an archaeologist). The wife of Bader (she is Russian) arrived with her five-year-old son and with all their belongings. They settled into one of the barracks, she is going to work in the quality control laboratory as a laboratory assistant. The wife of Zeybold has also arrived (she is also Russian). She works in our dining room… As a chef (!). My friend Carl Fridrikhsen is also expecting the arrival of his wife from Kazakhstan (!) and also expects this to be for 'ever'. However, it is a 10–15 day journey by train in the current conditions from there (normally it only takes six days). She is staying there alone for now (in the Pavlodar region). The sister of Carl Germanovich was mobilized (working duty), and now she works in Turinsk (not far from Nizhni Tagil). The mother of C.G.'s wife returned back to Nal'chik (in the northern Caucasus) with her other daughter and granddaughter. His wife has already been given leave from her job (working as a bookkeeper in the Belgachskaya tractor station) under the pretext of moving to become a teacher (she is a biologist, she taught biology and chemistry in the school there). Carl sent her a permit from the construction crew to the local police station with the request to issue passage to Nizhni Tagil to bring (?) an outfit to her mobilized husband. Furthermore, Carl sent her (in the middle of July) a telegram about the need for her urgent departure due to his serious (?) illness.

But what about you, Lidusya? Are you really unable to visit? I think on the 6th of August I will 'persuade' you, and we will arrange the date of your arrival. Undertake a mission to the brick plant of the NKVD *Tagilstroy*, the Nizhni Tagil city. By the way, let the UII office issues you with 'travelling ration coupons' (these were introduced from the 1st of August). Ella Kromer buys bread on these cards in Nizhni Tagil, and she also managed to get *brynza* using the meat ration card. This is much more convenient than carrying a loaf on your person. Do not take pillows and blankets. We have plenty in the room. I will give you mine.

A few words about myself: Lisa has already received your money transfer (400 roubles), as I learned on the telephone. I will visit her soon. Recently I received 60 roubles (!) which is 20% of my May wages. I have 16–34 roubles on my current account. I wrote a report to director with a request to transfer 800 roubles (50%) to my family. The director decided to 'allow the transfer'.

I have a great 'selfish' happiness: it is the third day that I have received 800 g (!) of bread per day. Now I will only receive 550 g on every sixth day, since they provided only five rations of 800 g for six workers in the quality control service.

I will end here as at 6.00 a.m. the siren starts the reveille, and I intend to go hurriedly into the forest in search of mushrooms so that I can return home sooner and will have some time to sleep before the next night shift. I've been hunting for mushrooms twice already. The day before yesterday I went alone and collected a full basket of blue and yellow boletuses in four hours. I clean and eat the mushrooms together with Carl. Next week he will be on the night shift, so he will be going into the forest, and we will eat them, the two of us together. I buy many green vegetables (for vitamins): green onion, horseradish, beet tops. One time Carl and I even managed to buy a little potato for 60 roubles, and a little mushroom soup.

Lidusya, please write to me with more information about yourself. I still believe my observation about being a 'forgotten husband' has a certain basis in truth (?). I have only received a single postcard, which is not enough for me!

Letter 70

Nizhni Tagil,
Tuesday 10th August 1943

Dear Lidusenka!

I am writing to you with a pencil for the first time and in a completely unusual situation. I am sitting on a clean bunk in front of an open window. Heaps of trunks, bags and mattresses lie all around me. I can see flashes of lightning through the window, it is thundering, and the rain falls in heavy showers. We are undergoing the disinfection of the ninth barrack from fleas and other parasites — which is no small matter. Today at 6.00 a.m. the entire ninth barrack began to move out their belongings into the 'zone' area (i.e. the open yard), and then began the fumigation of the fleas with sulphur dioxide which is carried out in all of the rooms by burning sulphur according to the reaction $S + O_2 = SO_2$. Everything was quiet up until 6.00 p.m. (precisely during supper, in fact), when a thunderstorm unexpectedly broke out and a wall of black cloud grew above the forest. At the same time as this was happening our chief scolded her subordinates (and me most of all, of course) for our failure to overcome many defects in the plant's production. The appearance of the thunderclouds interrupted our 'pleasant' conversation thankfully, and all the quality control inspectors dashed back to their barracks to save their belongings. Recently the disinfection of the tenth barrack has finished, the windows were taken out, and it was aired. After grabbing my things I jumped through the first window I encountered. Here I sit, now alone near my possessions, and wait for my comrades to bring me supper and 800 g of bread. I receive 800 g of bread every five days, the same as the other quality control inspectors, and on one day I get 550 g since they only allocated five 800 g rations for six people. Therefore, we have made a rota so that the 550 g ration is to be received in turn. These days I eat the whole 800 g in one go during the evening after picking up the bad habits of my roommates. We fetch the bread before going to supper. We then eat supper without any bread (skilly with the oats) and then we go 'home'. We then seat ourselves harmoniously around a table, and, armed with our knives, begin to slice up the bread rations. A bundle of greens serves us as dessert in this case, as well as some horseradish and beet tops. I imagine, Lidusya, that you do not feel such a keen hunger for the whole day as I do; possibly because after eating such a concentrated intake of food I can be sure that there will be not one piece of bread until the evening.

By the way, I must remark with some regret that SO_2 does not have the same effect on the fleas that our medical unit dreamed of. I look at the wall, and to my horror I can see that it is infested with fleas, which, scenting human flesh after several days of enforced starvation (during the disinfection), have formed a friendly crowd among themselves in which to hunt…

After this small lyrical digression, one additional parenthetic remark — I have continued this letter in another room, the department of dryers, where we have managed to store our belongings with the kind permission of Wolf, the master of the drying shop (I knew him in Sverdlovsk). I ate my bread ration and settled down for the night, when suddenly a soldier brought me the letter that you had begun on the 31st of July and finished on the 3rd of August, the one that you told me about on the phone on the 6th of August.

Lidusya, you invite me round to visit (!). But I am very busy with my work at the brick plant, so that I am afraid that I will hardly be able to come (!). It is sad, but it's a fact. It is apparent that we will not meet this year. But this is not my main worry. How you will cope with the next winter? Apartment? Firewood? Potatoes? Vegetables? You have to overcome so many concerns and difficulties on your own. Plus you have your job to think about, you worry about your husband (!) and daughter (!!). Dear daughter! She will return from the dacha in 15 days tanned, a little more grown up, happy! New features of her character will have appeared… Please tell me all about all this. I have already become reconciled to the thought that you will not come to visit me. After the telephone conversation on the 6th of August I still had hope, but after receiving your letter today I see that the situation is actually so dire that it is unfair of me to plead any longer. The main thing now is to put all our efforts in to surviving through one additional winter of war — the second winter spent apart during this war. I am certain in my belief that it will be the last wartime winter and that we will live together by next summer. Good or bad, in honour or in disgrace — this is not so important. It is only important that I will strive to live with you in harmony, so that the winds of fate cannot destroy the harmony between us.

I embrace you in my mind, Lidusya. Strong kisses to my daughter. I believe you will successfully overcome all and any difficulties you may face during this coming winter by your energy, efficiency and thrift.

Your Armin.

Letter 71[105]

Tuesday 17th August 1943

Dear Lidusenka!

My dear wife! Three more days of our enforced separation have passed once more… (As I was writing this to you at 11.00 p.m. I wanted to sleep so badly that I fell asleep right here, sat at the table in the quality control office.)

Wednesday 18th August 1943, 11.00 p.m.

I am continuing this letter sat, once again, in the quality control office. First I will convey my current news to you. For sure, the first piece of news is about … food (!).

The evening bread ration on the 14th of August was only distributed in the morning in quantities of 600 g, and the same amount on the evening of the 15th and so on throughout the subsequent days. They have promised us that, after the 25th, it will return to the 800 g ration once more. From the 15th, they have ceased to serve the second course at dinner. Three times a day they serve up a soup containing only oats, potato, and a few pieces of fat; occasionally some fish or kvass is dished out in the evening. I could not get a second ration from G., and so he has returned 80 roubles back to me. Furthermore, I have sold my ration of bread, 600 g, yesterday evening (the 17th) in the dining room for a further 70 roubles.

I will give you a brief account of the state of my finances.

Protected fund	—	900 roubles
Funds for current expenditures	—	106 roubles
Borrowings	—	100 roubles
Altogether as of the 19th of August, 1.00 a.m.	—	1106 roubles

[105]Between the 12th and 14th of August 1943, Armin's wife, Lydia, succeeded in visiting him in Nizhni Tagil — EAZ.

My outgoings for the period of the 15th–18th of August were 168 roubles. Namely:

For a photograph (additionally)	—	50 roubles
For a pot	—	50 roubles
For three horseradishes	—	18 roubles
Milk, onions, and potatoes from Panya	—	50 roubles

These expenses require some explanation.

I sent to you the photos. Fuchs charged 80 roubles for 5 prints. (I paid 30 roubles to the commandant and 50 roubles to Fuchs). Furthermore, he cut off the piece of the negative with our faces on it, so that it will be possible to print more when needed. Do you like this photo? To my mind it came out well. True, because of the unnecessary retouching your face looks too thin and flat, and on conversely I came out looking fatter than in real life. I have sent you one photo thus far. I am planning to distribute the others to your mum, to my mum, to Tosha, and to Modestus. For sure, in this case not one photo will be left for me to have. Therefore, perhaps I will not give a photo to Modestus yet. If and when it is possible to send you the negative then he will be able to print it for himself. Write me your thoughts on this, and is it better to send these photos first to you or directly to the designated persons?

My outgoings to Panya are connected to my trips into the forest in search of mushrooms (twice I have been on these missions together with P. F. Kromer) and the subsequent cooking of these on the bonfire behind Panya's house. Both times the mushroom dishes turned out to be first-rate, and I greatly regret that we could not share this meal with you. Did you receive the money transfer of 800 roubles? I enquired after this in the accounts office and they told me 'your money is on its way'. Let this money serve as a small compensation for the help you have given me (!).

How was your journey home, Lidusya? Did you visit the director to talk about the apartment? How have your housing problems been resolved? You will receive this letter probably on the 26th or 27th of August. Our daughter will be at home already. Please tell me about your impressions of her. Certainly, she will have changed after a month of staying on the dacha. Was your departure to Nizhni Tagil noticed with respect to the Institute, classes, and courses?

I will phone you on the 31st of August (Tuesday) at 9.30 p.m. Try to be at Modestus' flat on that evening. Kisses to you and my daughter.

Your Armin.

[Written in the margin:]

The days without you stretch out in boredom again! I often think about those three days that we spent together!

Letter 72

Nizhni Tagil,
Saturday 28th August 1943

My dear Lidusenka!!

I received your letter yesterday, the one you wrote on Sunday the 15th of August at Tosha's on the day that you returned from Nizhni Tagil to Sverdlovsk. It is good that you left on the night from Saturday to Sunday. You were able to rest on the train, and only pleasant recollections remain from the trip. I am also full of excitement over those three strange warm sunny days of August (the 12th, 13th, 14th) that we spent *together*, with you in Panya's cottage, in the forest, on the lawn.

Remember, on the last day in the evening we rested in this meadow encircled with trees (not far from Panya's house) and hundreds of flying ants were 'getting married'. Several moments of joint happiness and then fate separated them, and then they flew away in opposite directions. You and I, we are like these flying ants, we were fated to stay together for only three days during this year and then to depart again for another entire year.

I re-read your letters before and after your visit. What a difference in your mood! This trip cheered you up! You heaped bitter reproaches on me in the letter written before your arrival for my insistence that you visit me. And was I right? It would have been difficult to live apart for one more winter (and maybe even another summer?!) if we had not met in August! And everything turned out well. Nothing darkened these past three days. The weather was warm and sunny, and they let me take time off from my job; there were no complications with the mission certificate and the return ticket; and I was even able to see you off at the train station (!). Yet, soon after your departure the weather turned nasty. 'Autumnal light rain' showered down on us for a whole week (since the 20th of August it has been cold and sad; the cold winter returns once more).

Last night, at 11.00 p.m., Carl fell ill and went home to the barracks, so I was left to work on my own today (instead of the usual three of us!) as a quality control inspector + a foreman + a quality inspector for the entire plant during the night shift. There is a saw that is making a lot of noise in the next room. This is the work of Maria Trofimovna sawing bricks prior to testing them for their compressive strength. It is windy, cold and raining outside. The loading of railroad cars is ongoing; it is necessary to go to write out permits to transport the bricks... But my thoughts are not on the job in hand. My thoughts are with you, with our daughter, in Sverdlovsk, in the UBAS... Now Elza has returned from the dacha and you are together with her again. Although it is wearisome to care for her, kiddies can give us much happiness. I think about my daughter very often, I try to visualize her from the photograph now that she is one and a half years older than last I saw her. Lidusya, please describe to me in detail the first moments after Elza arrived back from the dacha. Was she happy to be back home? Did she feel disappointed about having to leave the dacha? Did she get upset because you came to visit me without her? What did she ask about me? Dear, glorious daughter...

Two weeks have passed since you and I plunged back into the whirlpool (or into the swamp?) of ordinary life. You — into the current, and I — into the swamp... I visualize the whirlpool twisting: classes with students, the installation of some new laboratory equipment, troubles with the apartment, other troubles (stealing from the vegetable garden, rats eating the last pieces of food that you have attempted to save for our daughter), scurrying between the dining rooms, the stores, hurrying your daughter along to the kindergarten, your domestic concerns (washing, darning, etc.).

Nevertheless, I ask you, Lidusya, to search for moments (in the dining room, at the tram stop) to inform me about your life. Did you receive the photo? I sent another three photos to Appolinaria Fedorovna, to my mum, to Elga. The fifth photo I kept for myself. If Tosha wants to have this photo, then I will send mine to her, and Fuchs will print one more.

Strongly, strongly kiss you, my dear wife.

Your Armin.

Letter 73

Nizhni Tagil City,
Tuesday 31st August 1943

Lidusenka, darling!

At this point two weeks have passed already since your departure! These two weeks have passed in some confusion. I have worked continuously on the night shift during the third week, but every day it is necessary either to stay after work in the morning or to start work earlier in the evening, for various reasons. However, there have been some 'achievements'. First, three days ago I had my teeth treated by our 'brick works' dental doctor. He put in place two new cement fillings to replace two that were lost. Moreover, he made these fillings in a suitably simple manner. In more detail: he cleaned out inside the cavities in each tooth with a hook, and removed a large mass of left-over food (!). Then he filled in the cavity with cement without using any disinfection (neither using ether, alcohol nor with any hot air?) or needing to do any further drilling. Half an hour after allowing the second filling to set, half of it fell out again, and when I returned to him with this piece of filling in my hand our doctor, being in some confusion, stated that this must be an 'excess piece' of filling. Of course, I did not believe him, but I am comforted by the fact that the dispensary is next to our quality control office, and I can renew this filling at any moment. I will hope that this simplified method of setting fillings will not cause the teeth to rot further underneath the filling.

The second achievement is that for the next five-day week I will be receiving 800 g of bread a day without the 'penal' 550 g ration as they have 'taken away' one quality control inspector and appointed him as the chief of the mining department. True, it was necessary to bustle about and to run a little in order to achieve this success for myself, since this extra effort was needed so that I would be included on the list of those to receive 800 g of bread instead of this other quality control inspector. But the food office did not want to do this without an order from the director. So yesterday (the 30th of August) this was finally solved. Our lady-in-chief obtained a permit from the director, and now I will receive 800 g during the forthcoming five-day week (until the 3rd of September). The only question is for how long will this remain the case?

Apart from these achievements there have been a few mishaps. First, I have scraped my face (on the temple and cheeks) and my hands whilst on duty, after stumbling over the feet of a sleeping (!) stoker in the complete darkness (there is a dearth of lamps) surrounding the calcining furnace. There have been many sarcastic remarks and laughter made in regard to this. But at first I was in no mood for laughter, when all the abrasions were hurting, and a huge bruise appeared above my right eye. But now it is funny to look back on this incident. Incidentally, I have made one 'scientific discovery' from this case. Namely: the expression 'sparks fell from the eyes' ['to see stars'] has a real meaning. When I struck my temple against the furnace then my first impressions were of a bright flash flickering in front of my eyes. Obviously, this is a physiological phenomenon from the shaking of the eyes or brains. Ask Tosha about this.

The second mishap was the fact that I had to combine two posts during the week, since the foreman overseeing the finished products was taken off to Zaygora to write posters. In connection with this it was therefore necessary for me to mark the bricks, to control their loading, to write the dockets, to select bricks and to send them for testing... Generally, it was necessary to remember my previous job. But the nights were rainy and cold. Brrr... For three nights I was required to do the work of three positions, since Fridrikhsen, a master of the face and the moulding was ill. Furthermore, one night (yesterday) I 'worked' in the place of four men: a quality control inspector of the kilning process, a quality control inspector of the mouldings, a foreman for quality inspection of the final product and still I worked as a laboratory assistant. Measurements of the surrounding air temperature were also needed (!), since a frost was expected and they were afraid that the raw materials would freeze in the sheds. The day after tomorrow I will go onto the day shift, but this does not hold the promise of much peace for me, as four 'apprentices' have arrived from the Molotovsky municipal construction technical school, and I have to train them in brick production.

This is how my working life has been spent since your departure, Lidusya! And I think more often than not that your arrival was very successful both in the sense of the weather and in the sense of my 'overloading' with work.

Lidusya, I wanted to have a phone conversation with you on the evening of the 31st of August, but alas! It turns out that all three municipal telephones at the plant are out of order, and so I must put off this wish for the time being.

Lidusenka, I received a letter from you dated the 15th of August, a postcard dated the 17th and a letter dated the 20th. I give thanks that you have not forgotten me. Now I await the next letter from you with a description of Elza returning from the dacha.

It would be good if they could give you some days off in September. Although you would not get much rest. It will be a change for you to gather in the vegetable garden without having to rush and to sort out the apartment and matters concerning the firewood. Please write to me about the steps you have undertaken in these areas.

Noskov has been granted six days of leave. It will be interesting to see whether UBAS can protect him or not. In any case, it will not be too bad even if he were to be forced to go into the Saratov region.

I end here. The time is approaching 1.00 a.m. It is time to go for the night supper. I am so grateful after your arrival, Lidusya! I am full of recollections about those days! Will we really not be able to see each other for one year more? I hope to see you sooner!

To a speedy (?) rendezvous, Armin.

[Written in the margin:]

Did you receive the 800 roubles? The accountant said that the transfers to our families had been sent already.

Letter 74[106]

Nizhni Tagil City,
Tuesday 7th September 1943

Dear Lidusenka!

So, we managed to have a phone conversation on the 3rd of September. It was so pleasant to hear your voice. It was pleasant to hear your stories about our daughter. Probably, she was glad when you went after her to the dacha at Shartash lake! Probably she chattered a great deal along the way from Shartash to Vtuzgorodok! I wish I could have been with you at that time (and not only at that time) to listen to Elza's chatter and to feel the entire family together… Recent events, our victories at the front, the disembarkation of the Allies in Italy hint at the hope that the hour of retribution approaches for Hitler, the end of the war approaches, and our meeting hour approaches too…

Markert visited me recently. I gave him a bucket of potato for repairing the rubber boots of Shpure (my schoolmate), and furthermore, he promised to repair my rubber boots for my commerce as well (!). He works in our garage repairing automobile tyres, while his speciality is as a teacher of physics. We chatted. He has had some family drama. His wife sent a letter requesting a divorce (!). She was his student in the school. After they were married it turned out that she was completely different from what he had thought her to be. They could not get on. I described how we met and were married. I recalled that I proposed to you only two months after the first time I saw you in the Ural Physico-Technical Institute (!). You were astonished. Then we had a trip on a steamship. (The lights were just turned off and so I am going to sleep.)

I take this letter up again on the 8th of September during the daytime in the quality control office. I want to describe to you how I managed to earn the potato that got my rubber boots repaired for free.

The fact is that my school mate Shpure promised me a bucket of potato in return for a favour (!) in May of… 1942(!), and so at that time I gave him my knapsack. Up until August 1943 he resided on the Lanina mountain

---

[106]The last from 1943 — EAZ.

(10 km from the brick plant), and in August they transferred him to yet another place, and I gave up any hope of receiving any potatoes in return, but I at least wanted my knapsack back. Then suddenly, on the 4th of September, he came to me and said that on the 5th of September he will be at the Lanina mountain and that if I wanted to pick up my knapsack and the potatoes, then I must be there on the 5th at 11.00 a.m. Luckily, a motor vehicle from the brick plant went to pick up some cabbage on the morning of 5th, and I arrived at the Lanina mountain at 11.00 a.m. However, Shpure was not in the barracks. After waiting for him for one hour I went to the house where his wife lived during the summer, and learned that he was in the vegetable garden, which is … one and a half hours away by foot! Fortunately, two of his neighbours were going in the same direction, so I left with them and managed to find Shpure in the allotment. It turned out that he had not had enough time to harvest the potatoes due to the morning rain. So I took a shovel and harvested two buckets of potatoes — one for myself and he asked me to take the second to Markert to pay for the repair of his boots. I boiled nearly half the bucket of potatoes on a bonfire and ate them heartily… At 4.00 p.m. I returned to the plant, and furthermore, Shpure persuaded me to go by a short cut — 5 km instead of 10 km. Soon I arrived at an old bridge, which, according to the directions Schpure had given me, I was not to cross. However, where the road continued past the bridge I could see that it soon disappeared under water, and, in spite of my searching, I could not find where it continued. I should mention that I was carrying more than 1.5 pounds as, apart from the two and a half buckets of potatoes in my knapsack I was also carrying about 6–8 kg of other vegetables that I had acquired along the road. In the end I reluctantly made my way across the old bridge (I did not want to return back the way I had come), and I carried on a little further. Within the hour, and according to Schpure's directions, I arrived at another much newer bridge, which was found to be in a more passable condition. This cheered me up a lot, so I crossed the bridge and asked a fat person whether the brick plant was much further away (according to my directions the brick plant should be very close to the new bridge). The man looked on me suspiciously and began to ask where I had gotten the potatoes. It turned out that I had walked in a complete circle and had returned once again to Lanina mountain — I had crossed the new bridge in the direction FROM the brick plant. That's that! I therefore had to walk for

another 10 km through terrible mud and the previous 5–6 km that I had walked counted for nothing. I dragged myself, hardly able to breathe, back to the brick plant and only arrived at 8.00 p.m. That is the amusing adventure that happened to me on Sunday the 5th of September, and despite this I remain very contented. First, I got my knapsack back along with the potatoes and some vegetables. Second, I got my boots repaired free of charge and jumped the queue. Third, I recalled my excursions and journeys from my childhood. In a word, I shook myself and felt that I was still alive. I found that my strength had not left me during these years of war. I am not so old yet!! Although I will be *33 years old* in eight days (on the 16th of September). My father was no longer alive at this age. He perished in the last war. And now there is a war again, and I am still alive.

Again they interrupt my writing. Vulakh came and it was necessary to construct the 'totals graphs'. I continue this letter later in the evening of the 8th of September at home. Recently a political instructor came to us with the information that Stalino City[107] is now occupied by our troops, and the Donbas[108] has been liberated in its entirety. It is good that these events have happened so quickly. My hope for a rapid end to the war, for the end of our forced separation, now grows with each day that passes.

I often recall those three days that I spent with you in the middle of August. The day before yesterday the wife of Paul Emilevich arrived (who suggested that we stay at Panya's house). They are also now living with Panya. Nadya stayed with Paul for five (!) days. Of course, Paul cannot be found either at home or at work; I meet him only in the dining room, shining and happy.

Five (!) days have passed since I last received a letter from you (the last postcard is dated the 23rd of August). Already I had become accustomed to receiving a letter every three to four days. Apparently, it is now necessary to break this habit again! Nevertheless, I wait your letter with a detailed description of our daughter's arrival, her stories about her life at

---

[107]Stalino City (Сталино) is now the modern-day city of Donetsk in Eastern Ukraine. Its former name derived not from Joseph Stalin, as Stalingrad was named after, but from the Russian word 'Стал' pronounced 'Stal', meaning 'steel' — GGW.
[108]The Donets Basin — GGW.

Shartash Lake. What news of the apartment, the allotment, the firewood? It is a shame that I cannot help you to solve all these 'cursed' questions.

Your HUSBAND.

Conclusion

Ten days after writing his last letter to his wife from the GULAG, Armin was demobilized from the brick plant at the Nizhni Tagil camp and directed to work at the Sverdlovsk Industrial Institute 'to be used according to his profession'. The demobilization certificate (pictured below) reads:

> *'To the Director of the Sverdlovsk Industrial Institute, Sverdlovsk City,*
>
> *On the orders of the Deputy Director of the GULAG NKVD USSR, Commissar of State Security Dobrynin, a demobilized soldier STROMBERG Armin Genrihovich is directed, to be used according to his profession.*
>
> *Signed: the Head of the Second Department of the Tagil, LAG NKVD, Loginov.'*[109]

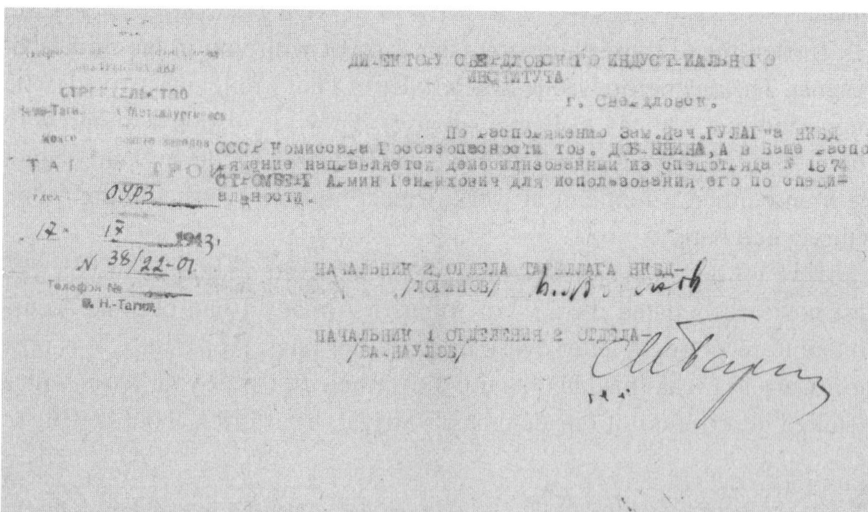
№ 38/22-01
1943
НАЧАЛЬНИК 1 ОТДЕЛЕНИЯ 2 ОТДЕЛА

[109]The 'Second Department' was the name of the record-keeping office that worked directly to serve the needs of the army in every state organization.

The details of exactly how Armin's release was brought about remain unclear. Armin later wrote in his memoirs:

> *'In September of 1943 I was directed back to the same institute from whence I was first mobilized, with a curious instructive document. [...] To whom I am obliged for making my release possible I do not know. This was, of course, an exceptional case since they freed nobody else, not even B.V. Rauschenbach.'*

There is some speculation that Armin's mother, Magda, who was still greatly respected in both scientific and society circles, was able to use her connections and influence on the rector of the Sverdlovsk Industrial Institute to plead her case at the highest levels, and Dobrynin himself was rumoured to be amenable to bribes. In any event, Armin's release was timely given that those remaining when the camps closed at the end of the war in 1945 had to remain in the closed city of Nizhni Tagil, without any possibility of travelling outside of the city, for a further ten years.

Thus Armin was able to return to his family in Sverdlovsk and to continue his work in the newly developing field of polarography. A whole new chapter in Armin's life was about to begin as his scientific career in his chosen field of specialization began to flourish both in his own laboratories and across the USSR as a whole.

4

Release, Work in Sverdlovsk and the Start of Polarography in the USSR

Introduction

Before we can examine Armin's life and scientific career following his release from the GULAG, it is instructive to briefly diverge from the biographical narrative to examine more closely the birth of his chosen field of specialization — electroanalysis using the technique of polarography — and to thereby better comprehend his later contributions to this important field of science. The Nobel Laureate, Jaroslav Heyrovský, pioneered the polarographic technique and his crowning moment of glory is a fitting place to begin the tale of a scientific method that would revolutionize analytical chemistry and profoundly influence the future life and career of Armin Stromberg.

The 1959 Nobel Prize for Chemistry Goes to Electroanalysis

On the 12th of December 1959 in Stockholm, a frail looking 68-year-old man in poor health, suffering from circulatory and digestive illnesses, dressed in a dinner jacket and white bow tie stood nervously before King Gustav Adolf VI and his queen. The Czech scientist Jaroslav Heyrovský was about to become the recipient of that year's Nobel Prize in Chemistry. As with all such awards, a Presentation Speech is traditionally made before the award is conferred on the recipient by a member of the relevant

Nobel Committee for Chemistry of the Royal Swedish Academy of Sciences, the committee that has decided which of the various nominated scientists should win the prize for that year. On this occasion the speech was delivered by Professor A. Ölander[1]:

> *'Your Majesties, Royal Highnesses, Ladies and Gentlemen.*
>
> *Analytical chemistry is a science, fundamental not only to the other branches of scientific chemical research, but also to applied chemistry, the chemical industry. Further, it is important for the other natural sciences, both within inorganic and organic nature, for medical research, and for many humanistic and even jurisprudential sciences.*
>
> *The striving of analysts is not only to develop methods yielding accurate results, but even more important for practical work is that the analysis can be carried out rapidly, that it can be done using as small samples as possible, and that very small percentages of various substances can be detected and ascertained.*
>
> *Polarography is one of these micro-methods which are available to the modern analyst. Professor Bohumil Kucera of Prague once suggested to the young Jaroslav Heyrovský that he should study certain irregularities in connection with the capillarity of mercury and attempt to disclose their origin. This was one of the innumerable small problems constituting science. Heyrovský let the mercury flow through a glass capillary and weighed the drops. It was a slow and tedious method, and he resolved instead to measure the electric current obtained when he put a tension between the mercury in the capillary and that collecting at the bottom. The glass capillary does not terminate in the air, but in a solution, through which the current will now flow.*
>
> *Heyrovský found that this device could be used for something much more important than the original problem. It could be used for ascertaining very small quantities of the most diverse*

[1]http://nobelprize.org/nobel_prizes/chemistry/laureates/1959/press.html

substances dissolved in the water, and moreover, to measure their percentages.

Important new discoveries are found where they are not expected. Spectacular advances in our time have been achieved by great teams, and some people are saying that only teamwork is worthwhile, whereas a single scientist nowadays can do nothing. Well, teamwork can be efficient, and is organized when you have been able to set an aim for the work. But the new discoveries are made by some scientists who noticed something strange, possibly by two, independently and in different countries. Then it is important that future team leaders and authorities granting funds do not keep him strictly to attend to his job, but give him a chance to pick up the unexpected new things, in spite of the chance of finding them being diminutive. When Heyrovský put a small electric tension between the dropping mercury and that collecting at the bottom, he found in accordance with earlier experience that the current increased by steps when the tension was raised over certain fixed values. Earlier people usually introduced the current into the solution by means of a platinum foil. But various substances will stick to the surface of a solid, disturbing the course of the experiment. Because the mercury is falling off after a few seconds and a new one then will be forming, this method always ensures us a new clean surface against the solution, and disturbances are avoided.

The current will increase perceptibly even at the very small percentages of such substances that will undergo a chemical reaction at the surface of the mercury drop when the current passes. The increase will not be unlimited, but is proportional to the percentage. The tension is a characteristic of each substance, and therefore the method gives us information concerning both which substances are present in the solution and their quantities.

Heyrovský, together with his Japanese collaborator Shikata, built an apparatus which registered how these electric currents varied with the tension applied. The apparatus, named the

equipment that was sufficiently cheap as to allow the widespread uptake by industry.

The origins of Heyrovský's invention reflect both the internationality of scientific research even before the early part of the 20th century and also in more than one way the frequent role of chance and of the individual in 'big' scientific discoveries. As Zuman observes:

> *'As with many other discoveries, that of polarography happened because the right man was in the right place at the right time. The man was right, because he had a sufficient background in chemistry to be able to develop a new scientific idea, because he had the inquisitive nature and drive that gave him the incentive to work hard on the development of his ideas and to sacrifice other enjoyments to achieve his goal, and because he also had the ability to observe and distinguish important new phenomena from those that were peripheral.'*[3]

Heyrovský was born on the 20th of December 1890 and raised in Prague. His father was Professor of Roman Law at Charles University. As a child he loved music, especially that of Wagner, and developed into an accomplished pianist, as well as enjoying sports such as soccer, skiing, swimming and hiking. In 1901 he entered the Akademické Gymnasium (high school) in Prague, was taught Greek and Latin for five and eight years, respectively, and developed his initial scientific interests in the areas of botany, zoology, mineralogy and astronomy before focusing on physics, chemistry and mathematics for study in his freshman year of 1909 in the Faculty of Philosophy at Charles University. As a result he became intrigued with the then newly developing interdisciplinary area of physical chemistry and decided that this was the area on which to focus. However, courses on this emerging topic did not exist in Prague so the young scientist persuaded his father, by then the rector of Charles University, to allow him to study in the University of London at University College, matriculating in 1910.

At University College the dominant figure was Sir William Ramsay, a Glaswegian who had originally trained as an organic chemist in

---

[3]P. Zuman, *Analyst*, 117, (1992), 1803.

Germany before returning to the United Kingdom, first at Bristol and then at London, and becoming interested in physical and then inorganic chemistry. He was the fourth winner of the Nobel Prize in 1904 *'in recognition of his services in the discovery of the inert gaseous elements in air'* — the gases argon, helium, neon, krypton and xenon — and his collaborator, Lord Rayleigh received the Physics Prize in the same year for the discovery of argon.

Ramsay retired in 1913, the year in which Heyrovský graduated with his BSc degree from the University of London. His successor was the physical chemist F.G. Donnan who was an Irish electrochemist, famed for

A caricature of Sir William Ramsay from *Vanity Fair*.

his work on membranes and on ionic transport in biological cells. Donnan was an undergraduate at Queen's University, Belfast, before training with Wilhelm Ostwald, seen today as one of the modern founders of the field of physical chemistry, at the University of Leipzig (where Donnan's doctorate was awarded in 1896). He subsequently spent a period of post-doctoral research with another founder of physical chemistry — J.H. van't Hoff in Berlin — before accepting his first appointment to the academic staff at University College, London. From there he moved to become a lecturer at the Royal College of Science in Dublin in 1903, and finally took up the Chair of Physical Chemistry at Liverpool University in 1906 until his appointment as Ramsay's successor.

Heyrovský, having completed his BSc, decisively chose to start his postgraduate experimental research under the supervision of Donnan. The project concerned the determination of the electrode potential of aluminium. This is not straightforward since a passivating layer of oxide on the solid surface compromises the obvious approach of immersing an aluminium electrode into a solution containing aluminium ions. This layer inhibits the processes that would otherwise establish an equilibrium potential on the electrode. Moreover, the reduction of water to hydrogen can take place rather easily. So Heyrovský's early experiments with solid aluminium electrodes failed. Donnan intervened to suggest using a dilute solution of aluminium dissolved in mercury, a so-called 'amalgam'. He further suggested, building on his experience with membrane equilibria, in order to overcome the passivation problem that the mercury be allowed to flow out of a glass capillary so renewing the surface, hopefully before there was a chance for surface passivation effects to become established.

Although the flowing mercury approach solved the passivation problem it did not overcome the hydrogen evolution issue and this prevented useful measurements being made immediately. The experimental problem remained unsolved when World War I broke out in the summer of 1914. At the time, Heyrovský was visiting Prague and he was unable to return to London. Soon after, he was called up for military service in the Austrian-Hungarian Army, but avoided action through being posted as a dispensing chemist and X-ray technician first in Tábor, South Bohemia, and later near Innsbruck in Austria. Nevertheless, Heyrovský was able to use his time in the pharmacy to prepare his doctoral thesis,

The Electro-Affinity of Aluminium. This he submitted in 1918 to the Faculty of Philosophy at Charles University, Prague, and was awarded his PhD in September of that year.

The final examinations for the PhD degree included one in physics alongside a longer examination in chemistry. These were conducted orally, in the case of physics, by Professor B. Kučera who at the time was working extensively on 'electrocapillary curves' measuring the surface tension of mercury as a function of an applied electrical potential. The initial experiments in the area dated from the work of G. Lippmann in 1873 but Kuera had an improved experimental method which involved letting mercury drop out every few seconds or so from a glass capillary connected to a mercury reservoir — in effect the original 'dropping mercury electrode'! Kuera had noted discrepancies — in the form of unexpected discontinuous maxima in the curves recorded with his device — which were quite simply unobserved in the classical experiments of Lippmann. In the course of Heyrovský's examination, as in many doctoral examinations since, discussion moved to areas of the examiners interest with Kučera describing his results. Heyrovský's response must have impressed his examiner since he was immediately invited to work with him and his colleague Dr Šimnek on the understanding of these curious results.

Heyrovský and Šimnek would spend many hours over the next few years tediously measuring the weight of mercury that had dropped from their capillary and how this depended on the applied potential. They developed a short cut in which they monitored the lifetime of each drop to replace the weighing method of the original Kučera experiment, but the work made little substantive progress. In early 1919 Heyrovský was appointed to the position of 'reader' in the Department of Inorganic and Analytical Chemistry under the headship of Professor B. Brauner who was an expert in the chemistry of the lanthanide elements. Under his influence Heyrovský researched on the structure of the aluminates and the phenomenon of amphoteric chemistry whereby a species can act both as an acid and as a base resulting in several research papers in 1920 which formed the basis both for a habilitation thesis and for the degree of DSc from the University of London which was awarded in 1920. With these qualifications, later that year Heyrovský was appointed as the first Associate Professor ('Docent') in Physical Chemistry at Charles

University responsible for giving lectures in physical chemistry. The young man who had left Prague for London in 1910 fired by his enthusiasm to learn more about physical chemistry had returned to his alma mater and had ended up teaching the topic he loved.

Throughout this period of promotion and diversification the electrocapillary work had continued to occupy much of his time; he discovered that the presence in the aqueous solution of divalent ions such as those of zinc, barium, manganese and cadmium caused changes in the surface tension curves, which suggested to Heyrovský the possibility of measuring their 'decomposition voltages' but came to realize that the drop lifetime method was not well-suited for this task. The breakthrough came when in late 1921 he hit upon the idea of using the electrical current flowing between the drop and the pool of mercury gathering un-weighed in the bottom of the experimental cell. The set-up for the new experiment looked much like that shown in Figure 4.1 and taken from the research publications in which the work was ultimately described in the journals *Chemické Listy* (the original report in Czech) and *Philosophical Magazine* (a subsequent translation into English).[4] The latter describes the apparatus as follows:

> *'The Apparatus. A double lead accumulator, A, was circuited through two resistance boxes, R. From one of these the current was branched to the electrolyzing vessel, which consisted of a conical flask, B, with a platinum contact attached to the bottom; a thick-walled glass capillary passed through the cork stopper, C, connected by a rubber tube to a mercury reservoir (not shown in the sketch), by the level of which the rate of dropping could be regulated. The glass tubes, which passed through the stopper, served to lead the hydrogen in and out. The negative terminal of the resistance-box was connected to the mercury reservoir above the capillary, the other was connected through a sensitive galvanometer to the bottom mercury layer in the flask.*
>
> *The capillaries were drawn out of a thick-walled thermometer capillary, with about 1mm inner diameter; they were broken off under water when connected with the mercury reservoir, so that*

[4]J. Heyrovský, *Chemické Listy*, 16, (1922), 256; J. Heyrovský, *Phil. Mag.* 45, (1923), 303.

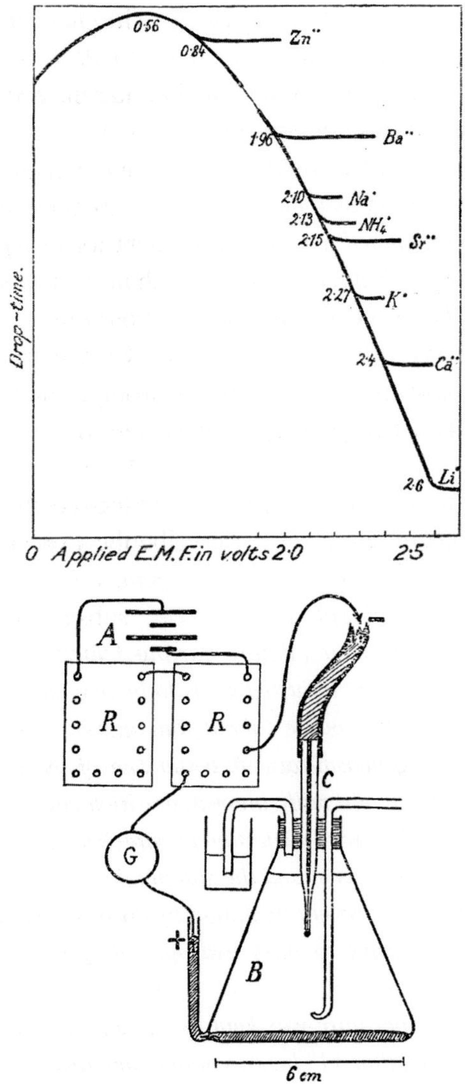

Figure 4.1: The surface tension vs potential curves (**top**) measured used the drop lifetime method rather than the classical Lippmann approach showed discontinuities in the presence of alkali and alkaline earth metal cations as indicated in this figure which also shows the experimental set-up (**bottom**) taken from J. Heyrovský. *Phil. Mag.* 45, (1923), 303.

> *the pure and dry mercury rushed into the tip first. It is essential to have the inside of the tube quite free from dust — however, the capillaries must not be washed nor moistened; for this reason capillaries made of fresh glass have to be used. The break of the tip was chosen so as to yield drops of about four to six seconds times (i.e. 0.02–0.05 mm, inner diameter of the opening).*
>
> *Since on prolonged use the capillaries get stopped and moistened inside, which might lead to hydrogen evolution in the tip, they have to be frequently exchanged for new ones.*
>
> *The reflecting galvanometer of the Kelvin type with an astatic needle had 50 ohms resistance; 1mm deflexion of scale, when 2m apart from the mirror, corresponded to 10^{-8} ampere.'*

The time-consuming nature of the research is apparent! The paper in the *Philosophical Magazine* goes on to describe the experimental method in detail:

> *'The Method of Procedure. Before starting polarization, the resistance of the boxes were adjusted so as to be one ohm per millivolt of the E.M.F. of the course (about four volts); this E.M.F. was determined by compensation with a standard Weston cell with an accuracy to a tenth of a millivolt and was frequently checked. The vessel was then filled with a solution prepared from Kahlbaum's or Merck's purest specimens, and the stopper was sealed with paraffin wax. In order to prevent oxidation by air at the cathode, it had been found necessary to pass hydrogen for 30 to 40 minutes through the solution.*
>
> *The room temperature was kept as close as possible to 20°C; the maximal deviation ±1° had no noticeable effect, which probably lies within the experimental errors of the method.*
>
> *The zero reading of the galvanometer was then taken and the polarization was gradually increased, and for each value of the applied E.M.F. the corresponding galvanometer reading was noted.'*

The first experiment of the current measuring type was attempted on the first day of January 1922 but was inclusive since the galvanometer used to measure the current was not sufficiently sensitive for the task. Heyrovský had insufficient funds to purchase the required equipment so it was more than a month later when the loan of a sensitive galvanometer had been arranged that the first 'polarogram' of aqueous sodium hydroxide was recorded. Zuman describes the events as follows.[5]

> *'On February 10, 1922, J. Heyrovský built a circuit consisting of a potentiometer, an electrolytic cell and the galvanometer. He placed a solution of 1M NaOH into the cell, immersed a dropping mercury electrode into this solution and added some metallic mercury to form a mercury pool electrode that was used as a reference electrode.[6] The current flowing between the dropping mercury and mercury pool electrodes was indicated by the galvanometer. Already, when a small voltage was applied, Heyrovský observed and recorded that the galvanometer indicated a weak current, the intensity of which oscillated rhythmically with the replacement of drops. With stepwise increase in voltage the current increased somewhat and at a more negative voltage the current increased again. These current increases corresponded to reduction of oxygen, which was understood in detail by Heyrovský only later. In the region between –1.9 and –2.0 V, Heyrovský observed on February 10, 1922, a substantial increase in current, corresponding to the deposition of alkali metal ions (Na^+) forming an amalgam, and that became the centre of his interest. Thanks to his background in electrochemistry, Heyrovský clearly recognized that he was on the track of an important scientific discovery.'*

Zuman, a former student of Heyrovský and who worked with him from 1948, continues to relate how the work progressed:

> *'During the following weeks his normally high intensity of work was raised to a feverish pitch. Every page of a 200-page thick*

[5]P. Zuman, *Critical Reviews in Analytical Chemistry*, 31, (2001), 281.

[6]The cell operated in modern parlance as a 'two-electrode' cell with the mercury pool acting both as a reference and counter electrode.

> *notebook was filled with laboratory notes during a period of seven weeks. One week after he recorded the first current–voltage curve he restricted his experimental work to obtaining such curves. Soon he realized that oxygen is reduced at the dropping mercury electrode (D.M.E.) in two steps[7] and also that the current due to oxygen reduction interferes with measurements of currents due to other processes. From the beginning of April, Heyrovský removed the dissolved oxygen (present in solutions in contact with air) by bubbling through the solution in the electrolytic cell a stream of hydrogen that he generated in a Kipp apparatus.'*

As we have seen, the results of these efforts were published first in Czech and then in English. These papers show the first ever polarograms in the scientific literature and are reproduced in Figure 4.2 where the significant difference between the current–voltage curves measured in air and under hydrogen is apparent, as are the different potentials required to form the

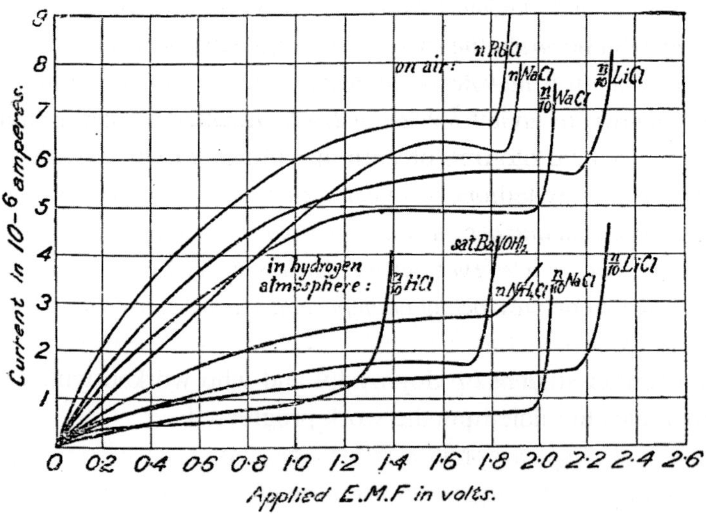

Figure 4.2: The first ever polarograms to be published in the scientific literature reproduced from J. Heyrovský *Phil. Mag.* 45, (1923), 303.

---

[7]The first of the two steps forms hydrogen peroxide, the second water.

amalgams of lithium, barium, sodium and rubidium from the corresponding alkali metal cations. These potentials were found to be concentration dependent and the values related to the thermodynamics of concentration cells whereby a tenfold change of concentration of a uni-valent and bi-valent cation produced characteristic changes of potential, differing by a factor of two. The start of a huge effort in the study and application of polarograms and current–voltage curves had begun.[8]

Polarography Revolutionizes Analytical Chemistry[9,10]

Heyrovský's scientific success led rapidly to promotion and in the same year as the two pioneering papers on polarography he became the Head of the Physical Chemistry Department at Charles University, newly established in 1922. This marked the start of a rapidly enlarging team devoted to polarographic studies. Scientific impact increased as a result of publications recognizing first the role of diffusion of the electroactive species in controlling the maximum current (the 'limiting current') that can be passed at the D.M.E. and in extending the work into solvents other than water. However, possibly the most significant achievement and one which did most to enhance the role of polarography as a general analytical tool was the invention of an automated instrument called a polarograph which replaced the manual plotting of current–voltage curves. Needless to say the latter was hugely tedious — with each plot requiring over an hour — whereas the new instrument allowed an entire polarographic curve to be recorded in a matter of a few minutes. Today, of course we assume almost every measurement will be made with automatic recording of the data. However, in the 1920s this was essentially unknown and so the invention was both sophisticated and pioneering. It is noteworthy that such automation predated the corresponding transformation of the analytically important field of spectrophotometry by around two decades!

---

[8]The term 'polarogram' usually signifies the measurement of a current–voltage curve at a D.M.E. In the context of other electrodes the term 'voltammogram' is commonly, but not exclusively, preferred.

[9]P. Zuman, *Critical Reviews in Analytical Chemistry*, 31, (2001), 281.

[10]P. Zuman, *Analyst*, 117, (1992), 1803.

Heyrovský made the invention of the polarograph in collaboration with a Japanese co-worker, Masuzo Shikata. The latter had graduated from the Department of Agricultural Chemistry in Tokyo University and moved to the Research Institute of Physics and Chemistry (also in Tokyo) where he developed an interest in electrochemistry. He undertook a research visit to the laboratories of Professor I. Traube in Berlin but whilst in Germany excitedly learned of Heyrovský's research in electroanalysis and decamped, joining the Prague team in 1923. His paper with Heyrovský describing the polarograph was published in 1925. The instrument (shown in Figure 4.3) recorded the polarogram photographically using the reflection of a beam of light from a galvanometer.

The automated polarograph allowed extremely rapid scientific progress. Of fundamental significance were experiments changing the internal radius of the capillary and the height of the mercury reservoir and thus the rate of flow of mercury. This was shown to control the size of the limiting current, I_{lim}, and the effects were quantified by means of the

Figure 4.3: The automated polarograph instrument invented by Shikata and Heyrovský.

Ilkovic equation[11] relating this to the diffusion coefficient (*D*, measured in cm^2s^{-1}), the rate of mercury flow (*m*, measured in mg s^{-1}), the drop-time (*t*, measured in s) and the concentration of the species undergoing electrolysis (*c*, mmol dm^{-3}):

$$I_{lim} = 607ncD^{1/2}m^{2/3}t^{1/6}$$

where n is the number of electrons transferred per molecule in the electrolysis. The equation is named after the scientist who derived it, the Slovak, Dionyz Ilkovic, a research assistant of Heyrovský in the 1930s.

The Ilkovic equation describes simple electrolysis at a D.M.E. to an excellent level of accuracy. By 'simple' is meant the case where a molecule diffuses to the mercury drop, undergoes electrolysis and then the products diffuse away from the mercury surface. It was soon recognized that the rate of chemical reactions in the solution could also influence the limiting current — so-called 'kinetic currents' which arise when the molecule that actually undergoes electrolysis at the D.M.E. is involved in chemical equilibria within the solution so that the rate at which it is generated controls the current as well as the rate of diffusion to the electrode. That said, regardless of the precise mechanism of electrolysis it was rapidly clear that because the electrical currents flowing through the cell could be measured with great sensitivity it was possible to use the magnitude of the limiting current analytically, that is to say to measure concentrations in the solution.

The analytical uptake of the method was stimulated by two major factors. First, Heyrovský and others were extremely energetic in promoting his work. His team published research papers prolifically, an activity aided by the founding of the journal *Collection of the Czechoslovak Chemical Communications* in 1929. He made visits abroad to lecture and participate in scientific conferences, most notably to the USA in 1933 where he spent six months lecturing at the University of California, Berkeley. Key textbooks were published: the book *Applications of Polarography in Practical Chemistry* appeared in Czech in 1933 and was translated and published in 1937 in Russian. G. Semarano published *Polarography, its Theory and*

[11] D. Ilkovic, *Collect. Czech. Chem. Comm.*, 6, (1934), 498; D. Ilkovic, *J. Chim. Phys.*, 35, (1938), 129.

Application in Italian in 1932 whilst a book chapter by Heyrovský himself appeared in the influential series *Physical Methods in Analytical Chemistry* edited in Germany by W. Bottger in 1939. Kolthoff and Lingane published *Polarography, Polarographic Analysis and Voltammetry, Amperometric Titrations* in New York in 1939 and had a major impact in North America. Also, Heyrovský's early students returned to their home countries to dissipate news of the new science and in the cases of Kemula (Poland) and Semerano (Italy) to establish influential schools of electroanalysis.

The second factor promoting the widespread use of analytical polarography was the availability of commercial equipment from as early as 1929 supplied initially by V. and J. Nejedlý of Prague and of the automatic photographic recording type. Subsequent manufacturers were Leybold in Germany and E.H. Sargent and Co in the USA, the latter holding a US patent for the device. Other companies followed — Cambridge Instruments (UK), Zbrojovka (Prague), Yanagimoto, Shimadzu (both in Japan) and Radiometer (Copenhagen). The photographic recording polarograph was manufactured up to about 1950 although pen recorders were introduced by Radiometer in 1938.

Polarography in its simple form — so-called 'd.c. polarography' — can measure concentrations as low as *ca.* 10^{-5}–10^{-6} moles per litre. Areas which took up the technique as a simple, cost-effective and sensitive approach included the pharmaceutical industry for quality control and for the investigation of the metabolism of drugs *in vivo*; the chemical and metallurgical industries for monitoring materials at all stages of the manufacturing process; and the food and synthetic rubber industries. The methodology would also impact the world of atomic energy research where the need for measuring even smaller levels of concentration than was possible with d.c. polarography would stimulate the invention of new techniques which, in the 1950s and early 1960s would transform electroanalysis.

Polarography Changes: Pulse Voltammetry, Static Mercury Drops and Stripping Voltammetry

On the 1st of January 1946 the United Kingdom's Atomic Energy Research Establishment was formed at the site of RAF Harwell about sixteen miles south of Oxford between Didcot and the village of Harwell. The incoming scientists simply took over buildings and accommodation from the departing airmen. AERE (or 'Harwell' as it simply became known) was tasked by the Ministry of Supply to research the use of nuclear fission for both military purposes and for energy generation. The site had been selected partly because of the need for good water supply and excellent transport links but also because of the proximity of a quality nuclear physics laboratory at the nearby Oxford University. Whilst Cambridge offered similar advantages, and arguably a superior nuclear physics facility, the RAF had been unwilling to relinquish any of its airfields in the east of the country because of the looming threat of the Cold War. The facility that was to become the UK's main centre for atomic energy research and development for approximately 50 years was therefore established in Oxfordshire.

The research at Harwell generated the need for analytical chemists to determine the course of the nuclear processes studied and, in the nature of the work, this was to provide a major challenge to existing methods of analysis. This focused stimulus led to developments under the leadership of G.C. Barker that greatly transformed polarography by increasing the analytical sensitivity of electroanalysis. He did this in two ways. First, the steadily increasing voltage applied to the electrode in the classical approach was replaced by one containing either a square waveform or a series of pulses. These give a greater perturbation of the electrolysis and hence a larger signal is available to be measured. Second, rather than measuring the current throughout the experiment, or in the case of classical poloarography, throughout the lifetime of each mercury drop, Barker realized that by postponing the measurement until late in the lifetime of the drop great sensitivity could be achieved. The reason for this is that the current due to electrolysis that contains the sought after analytical information is partially masked by other background currents caused by the increasing area of the mercury drop as it grows and the change in the applied potential. These background currents flow not as a result of electrolysis but as simply an electrostatic flow of ions, from the electrolyte in the cell, to or from the electrode surface as it becomes charged. Such 'capacitative currents' are largest just after the birth of each drop, and so by postponing the current measurements until they have almost died away an improved resolution of the analytical signal from the masking capacitative currents is achieved.

The result of Barker's improvements and the introduction of 'square wave voltammetry' and 'pulse polarography' was around a hundredfold increase in sensitivity; in favourable cases concentrations below 10^{-7} M, and approaching 10^{-8} M could be measured with the new techniques.

Towards the end of his 1952 paper (with I.L. Jenkins) in *The Analyst*,[12] having introduced square-wave voltammetry and demonstrated its much enhanced sensitivity, Barker speculates on the need, or otherwise, for further improvements and suggests how this might be possible.

> *'As regards any further improvement in sensitivity, it is believed that the useful concentration range of the present instrument might be extended to lower values by a factor of ten or more by paying more attention to the purity of chemicals, to the noise level of the electronic circuits and to the design of the capillary. The experimental technique would then become somewhat specialized and it may be doubted whether it would be acceptable for analytical purposes. Indeed there is some doubt in the authors' minds as to whether there is any real need for higher sensitivity. If there were, one might mention a recent development that can lead to a large increase in sensitivity for the detection of many of the species that form metallic amalgams. Briefly, this involves the use of a single mercury drop in place of the dropping mercury electrode, and a special cell is used that permits the circulation of the solution past the surface of the drop, the circulation system being such that the diffusion system in the vicinity of the drop can be reproduced. At the start of an experiment the drop electrode is polarized to as negative a potential as is possible without depositing the cations of the supporting electrolyte and is held at this potential for a suitable time. Metallic impurities in the solution then tend to be concentrated in the drop and, after 15 minutes have elapsed, the concentrations of the electro-deposited ions in the drop may well exceed their concentrations in the solution by a*

[12]G.C. Barker, I.L. Jenkins, *Analyst*, 920, (1952), 685.

factor of a hundred or more. After a suitable level of concentration has been effected, the circulation of the solution is stopped and a derivative polarogram is recorded with a relatively rapid rate of change of the mean potential of the drop.

If the conditions under which the experiment is carried out are correctly chosen, the heights of the waves observed on the polarogram are determined by the concentrations of the various metallic ions in the drop at the start of the polarogram, and these concentrations consequently can be determined from the measured wave heights. If the system is calibrated in some way, the concentrations in the drop can be used to estimate the original concentrations of the various ions in the solution. The accuracy of the method is probably not better than five to ten per cent, but the method is of interest as it is readily applicable to the estimation of concentrations as small as 10^{-9} M and, if pressed to its limit, one might expect the smallest amount of a single species that could be detected to be of the order of 10^{-11} moles. The method has been found useful for studying the purity of the supporting electrolyte.'

With this Barker is credited with the idea of pre-concentrating analytes on the electrode in order to lower detection limits and hence the invention of the technique now known as 'anodic stripping voltammetry'. The method was to prove extremely fruitful.

One of Heyrovský's early co-workers (1929–1930) was the Pole, Wiktor Kemula, who subsequently established a major centre of electroanalytical research in Warsaw. In 1958, Kemula with Kublik published a paper in *Analytica Chimica Acta*[1] in French entitled '*Application de la goutte pendante de mercure á la détermination de minims quantités de différents'*.[13] This paper described the construction of a hanging mercury drop electrode that could be used for electroanalysis. In other words, rather than the mercury periodically dropping out of a capillary, as in classical polarography, the Poles chose to work with a single drop, although of course this could be renewed from time to time. They used this device to illustrate the idea of Barker (cited as the first reference in the Polish

[13]W. Kemula and Z. Kublik, *Anal. Chim. Acta*, 18, (1958), 104.

paper) and showed how the hanging mercury drop could be used to estimate small quantities of aluminium and very small quantities of lithium as well as the detection of copper, thallium, lead, cadmium and zinc with just a few millilitres of solution needed for analysis.

Polarography and pulse and square wave voltammetry, along with stripping voltammetry, form the basis of modern electroanalysis and are the scientific legacy of Heyrovský and Barker. One of the many scientists to recognize their value at an early stage was Armin Stromberg.

Polarography Arrives in the USSR

In 1993 Stromberg wrote an article reflecting on 'half a century with polarography'. In it he explains how he first encountered the technique and its origins, from his perspective, of how it started to be used in the USSR[14]:

> *'It is more than 60 years since I learned about polarography. It was in 1932, when I read an article by Heyrovský dedicated to a decade of polarography.[15] It was a large review (around 40 pages in German), and it made a strong impression on me. At that time I was in my second year of working for the Ural Research Institute of Chemical Industry (UNICHIM, Sverdlovsk) as a fellow of the "magnesium group", immediately after graduating from the Urals Polytechnic Institute (UPI) in 1930 where I specialized in electrochemical engineering.*
>
> *Before World War II there were some papers on polarography in the USSR, but not very many. In my overview report on polarography (1951)[16] several publications were mentioned. Usatenko and Lyalikov (1936) proposed a methodology for determining iron in ore, slag and metallurgical production; copper and nickel in stainless steel (1937). Komar (1937) developed a methodology for*

[14]Translated from: A.G. Stromberg, *Russian Journal of Analytical Chemistry*, 48, (1993), 939.
[15]J. Heyrovský, *Microchemic.* 1932. Bd 12. S. 25–64.
[16]A.G. Stromberg, 'Works of the institute of chemistry and metallurgy of UBAS USSR'. 1951. Iss. 4, 30.

determining the amount of potassium and sodium in mineral sources, and Portnov and Afanasyev (1938) issued a method for the separate determination of potassium and sodium. Among the few works on the theory of polarography there was a series of works by Frumkin with co-authors on the theory of polarographic maxima, beginning with an article by Frumkin and Bruns in 1934. Kryukova published several articles and defended her doctoral dissertation on the verification of this theory (1939), but in general, as noted in the literature,[17] *before World War II polarography was developed mainly in Czechoslovakia, and only imperceptibly in other countries. It seems that this is due to the fact that new technology industries, primarily microelectronics and nuclear industry calling for determining trace [materials] at* 10^{-5}*–*10^{-3}*% and below, began to develop only after 1945 previously there was not such a large demand for this sensitivity.'*

Professor Herman Budnikov in a lecture in 2008,[18] addressing the development of voltammetry in Russia, considered that the development of polarographic analysis in the USSR began after the 1936 mission of academician Vladimir Vernadsky together with his colleague Alexander Vinogradov to a number of foreign countries and cities, including Prague, for the purpose of becoming acquainted with the state of the art of various methods of chemical analysis. Vernadsky was a polymath with interests in geochemistry, mineralogy, biogeochemistry and crystallography as well as being a most distinguished naturalist and philosopher, famous for introducing notions of the biosphere and ionosphere and the concept of life as a geological force that shapes the earth, although these ideas were not recognized in the West until much later. He was also one of the first scientists to recognize that the oxygen, nitrogen and carbon dioxide in the Earth's atmosphere result from biological processes. In 1928 he founded and was director of the Biogeochemical Laboratory of the USSR Academy of Sciences, which was located first in Leningrad, and then

[17] I. Kolthoff, J. Lingane, *Polarography.* Bd 1–2. 1952.

[18] Professor G.K. Budnikov, Kazan University, presented at the 7th Conference on Electrochemical Methods of Analysis, Ufa, Russia, June 2008.

transferred to Moscow. Later, in 1947, it was reorganized as the Vernadsky Institute of Geochemistry and Analytical Chemistry, Russian Academy of Science; this was, and is to this day, the leading institute for research on analytical chemistry in Russia.

It was Vernadsky's strong belief in the value of international collaboration that led him to travel extensively across Europe both for research and for teaching and took him and Vinogradov to Prague. The latter studied the new polarographic technique in Heyrovský's laboratory and returned to the Biogeochemical Laboratory to establish the technique there in 1936, reporting on his experiments in the First All-Union Conference on Analytical Chemistry held in late 1939 in a paper entitled 'The polarographic method in analytical chemistry'. In the same year courses on polarographic analysis were organized in Odessa by Professor Eugene Burkser and the first Soviet polarographs were produced in the same city correlating with the early reports of this type of analysis noted by Stromberg.

Vernadsky was also responsible for the Russian translation of Heyrovský's book *Applications of Polarography in Practical Chemistry*. Budnikov quotes from his diaries of 1937: 'The book was published on my initiative and was translated from the Czech by E.N. Varasova, a Czech woman who worked in my laboratory. I do not know where she is

Portrait and photograph of Vladimir Vernadsky (**left**) and Alexander Vinogradov (**right**), the pioneering missionaries of polarography in the USSR. Reproduced, with kind permission, from Prof. Budnikov's lecture slides.

now.' Varasova was the wife of USSR diplomat and had worked in Heyrovský's laboratory where she mastered the polarographic technique and published at least two papers in the *Collection of the Czechoslovak Chemical Communications,* one entitled 'Polarographic studies with the dropping mercury cathode: Part XVIII' and the other 'An investigation of soap solutions'. After returning to Leningrad and completing the translation with Vernadsky's encouragement she worked for a while in the State Institute of Applied Chemistry (Leningrad) but was sent to the camps where she perished.

The All-Union Conference on Analytical Chemistry of 1939 had, as well as Vinogradov's paper, at least ten other reports on polarography reflecting the fact that by the end of the 1930s the method had taken root in almost all branches of production: nonferrous metallurgy, the analysis of cast irons and steels, the analysis of minerals, medical and pharmaceutical analyses, and more.

Stromberg's personal reflection of this period is as follows[19]:

> *'For ten years after my graduation from university I dealt with different issues in the laboratory concerning the electrochemistry of molten salts in the Institute of Chemistry and Metallurgy, Ural branch of the USSR Academy of Sciences (ICHIM UFAN, Sverdlovsk). Polarography invaded my scientific and practical interests in 1939. That year the Institute received a polarograph from the Odessa pilot production workshops. In these workshops in 1936–1937 an analogue of the (1924) Heyrovský–Shikata polarograph was designed, and in 1938–1940 a series of several dozens of such devices was produced. Our laboratory got this device because we needed to control the purity of the salts used in our research. The head of the laboratory, S.V. Karpachev, ordered a new, young member of staff, Malyarovsky, to organize these controls.*
>
> *The polarograph from the Odessa workshop was a wooden box occupying about a quarter of a desk. An engine was inside, moving a rheostat slide wire and simultaneously rotating a hollow cylinder with a slit. The change of current was recorded on*

[19]Translated from A.G. Stromberg, *Russian Journal of Analytical Chemistry*, 48, (1993), 939.

photographic paper attached to the cylinder using a spot of light reflected from the mirror galvanometer, which was located on a main wall of a building, not prone to mechanical shaking. For its time it was the most perfect polarographic device.'

Stromberg goes on to relate how he subsequently came to have hands-on experience of the polarographic technique:

> *'In September of 1939 I defended my thesis*[20] *on the theory of viscous mixture of molten salts [...] I do not remember whose idea it was, but the Ural House of Technology proposed to me and one teacher from UPI to organize a school (laboratory classes) on polarography for factory workers on the premises of the Urals Department of Analytical Chemistry in UPI. We made five polarographs with reflecting mirror displays with the parts provided by the House of Technology, and which were constructed from two mechanical potentiometers and a mirror galvanometer suspended on the wall. Lead-acid batteries were the source of the current. Polarographic cells with mercury drop electrodes were also constructed and the second electrode was the mercury [in the cell] bottom. The voltage applied to the cell was changed manually in increments. A current value for each voltage was visually measured from the scale of the mirror galvanometer. We recorded them, and then plotted polarograms point-by-point. For analytical purposes the value of residual current and diffusion current were recorded. Hohn's book in German*[21] *was our guide. By training factory workers, I also studied the method. This was my first practical mastering of polarography.'*

Stromberg subsequently constructed polarographs for the Ural Mining and Metallurgical Technical School and, in 1941, introduced the apparatus into the Magnitogorsk Metallurgical Complex where it was used to determine micro-quantities of the alkali metals sodium and potassium.

[20]The thesis was that for the Candidate of Chemical Sciences, broadly equivalent to a modern-day PhD thesis from a British or American university.

[21]H. Hohn, *Chemische Analysen mit dem Polarographen*, Berlin, 1937.

Stromberg's Research in Sverdlovsk: Wartime and After

At the outbreak of war many of Stromberg's colleagues were mobilized and sent to fight at the front. By this time Stromberg held the position of Senior Scientific Worker and so, along with similarly highly qualified scientists, was able to remain at work, taking on the post of Manager of the Analytical Laboratory, working closely with polarography, which he successfully introduced into several Ural plants by means of instruments built in the machine shop of the institute. Stromberg reports that[22]:

> *'These were special-purpose analysers, which made it possible to measure the current only for the beginning and end of the wave for two–three selected element admixtures against the background. With the help of a special rheostat it was possible to regulate the height of the polarogram wave of the selected standard (reference) solution and to establish its required constant value (for example, one hundred centimetres according to the scale of the mirror galvanometer). This simplified the calculations of the element concentrations.'*

Plants benefitting from this instrumentation included Uralmash and the Middle-Urals Copper Foundry amongst others.[22]

> *'Primarily my work in these wartime months was to specify the conditions for the determination of impurities of cadmium and zinc in the products of the copper-smelting production and the introduction of these procedures at a number of plants, in particular at the Middle-Urals Copper Foundry, and also to repair polarographs, and to train the plant analysts to technology of polarographic analysis.*
>
> *It was impossible to do any serious studies in these months. It was necessary to store firewood for the winter, to harvest potatoes in a kolkhoz, to attend self-defence classes, etc. As a result, almost no time was left for science. No publications on polarographic analysis were realized at that time and I had no contacts with any*

---

[22]Translated from A.G. Stromberg, *Russian Journal of Analytical Chemistry*, 48, (1993), 939.

> *polarography developers. Apparently, I was the only polarography developer in Sverdlovsk (and maybe even the Urals).'*

In his overview of his life in polarography Stromberg only briefly covers his period of internment:

> *'In March 1942 I was formally mobilized into the army, and for one and a half years I was imprisoned into the NKVD's special-force camp for the Soviet Germans known as the GULAG, working in a brick plant in Nizhni Tagil town, because of my German birth. The conditions of stay in the camp were difficult; of six thousand Soviet Germans, who were in the construction crew, three thousand people died in three years. After the camp closed in 1945, all the inhabitants were forced to stay in Nizhni Tagil for more than ten years as deportees without the right to travel. Following the act of 5.10.91 entitled "On the rehabilitation of the victims of political repressions" all the surviving Germans were rehabilitated, and I was one of them, [with the formalities completed] in April 1992.*
>
> *In September 1943, thanks to a happy coincidence of circumstances I was released from the camp (after spending a year and a half in it), and with a covering letter I was sent back to the same institution "to work in accordance with my expertise". So, I continued to work in the institute, as the manager of the laboratory of analytical chemistry.'*

Elsewhere, Stromberg gives more details of the events leading up to his internment (March 1942) and of his release in September 1943 (see also Chapter 3).[23]

> *'At the end of 1941 a messenger dispatched a notice to me, in which it was instructed "to come next day to the police with the passports of your entire family". I already knew that this meant the moving of our entire family (including mother) as special-migrants into one of the northern settlements of the Urals. The best that could be hoped*

[23]A.G. Stromberg, '40 years of the Special Research Laboratory', TPU, 2004.

for [if this happened] would be work in the educational system as a teacher of chemistry. Such cases were already known to us in the first months after the beginning of war amongst some of the German-instructors in educational institutes. Fortunately, everything ended simply as a short, sharp shock. The same day [as I received the message] mother went to the Rector of the Ural Polytechnic University, and he resolved this issue, since he knew mother well and valued her as a person and as a worker. The fact is that the Rector was a promoted worker from the leading organs of the party and all the party leaders of the region knew him. He called whoever was necessary in order for us to be left in peace.

Several weeks later new trouble arose, which ended less satisfactorily. A messenger brought me a notice indicating that I "will be mobilized into the RKKA, [Workers' and Peasants' Red Army]" (I was exempt from joining the regular army as were all the scientific workers of the institute with a degree). I immediately visited the military registration and enlistment office where some other Germans had already collected. The military office staff warned us that we were to be mobilized into the "working army" and they ordered us to arrive the next day with our personal belongings. We travelled by railroad car and were brought to the NKVD's GULAG (German special-force 1874) in Nizhni Tagil town and to the brick plant which was surrounded with barbed wire.

The living conditions in this special-force were very hard. According to data from different sources, in the three years of its existence 3000 of 6000 German "labour-line soldiers" (about a half) died because of the hunger, cold and hard working conditions. I survived because soon after arrival into the camp it was possible to pass from the general works into the "administrative position" of the dispatcher of the calcining shop, then the master of quality control of this shop. Accordingly, from the room in the barrack where 40 people lived, I was moved into the room "for engineering-technical personnel", where eight people lived. I should mention at this point that in the same room lived B.V. Rauschenbach — the theoretical physicist, future academician, and later to be one of the organizers of space flights.

> *military themes were usually occupied by carefully selected members of [the communist] party). It therefore became urgently necessary for me to defend my doctoral degree. But I had no free scientific personnel for the necessary studies. So I took final-year students from the [Ural State] University, graduate students, including unofficial ones, and, against the rules temporarily transferred several colleagues to work on some theoretical studies.'*

In these eight years in IKHMI Stromberg managed nevertheless to publish 26 articles on polarography of which the first four appeared in 1945–1946. His retrospective description of the research shows the influence, and synergy, of both applied analytical need and the development of fundamental insights into the theory and application of polarography. His own account of these early papers is as follows:[26]

> *'The first article was dedicated to the polarographic determination of cobalt in the presence of nickel* [A.G. Stromberg and A.I. Zelyanskaya, Russian Journal of General Chemistry, 15, (1945), 426]. *The work is multi-faceted, both theoretical and practical questions of the polarographic analysis of cobalt are mentioned in it. The practical value of work lies in the fact that we found the method of determining cobalt in the presence of nickel. Waves of both elements are located very closely in different backgrounds, which hampers the determination of cobalt, for example in the ores, where it is always located together with nickel; moreover, the content of the latter usually is an order of magnitude higher. The method proposed lies in the to add a probe molecule, dimethylglyoxime, to an ammonium solution of the analyte. Then the complex of nickel falls out as a sediment, and cobalt remains in solution, but its diffusion wave is complicated by the catalytic wave of hydrogen (previously the catalytic wave of hydrogen in the presence of cobalt and cysteine was studied only by Brdichka in 1933). In my work the influence of different factors on the height and form of catalytic wave was studied, and a hypothesis about the mechanism of the process expressed. The procedure of the determination of cobalt in the ores was developed, and checked on*

model samples, obtained from the central laboratory of the Ural Geological Office and the Pyshma concentrating plant. The procedure proved to be simpler and more rapid than earlier ones that had been adapted for determination of cobalt. Subsequently our method was used in these organizations. I want to note that both in this and subsequent works of an applied nature I also tried to solve theoretical problems together alongside the practical analytical tasks. This combination made it possible for us to select the best conditions for analysis with the benefit of fundamental insights into the processes.

Later, together with colleagues of the central laboratory of Ural geological office V.A. Boretskaya and A.D. Naranovich, we developed another procedure for the polarographic determination of cobalt in ores, using the wave of the triple-charged ion of cobalt.

The use of dimethylglyoxime with the polarographic determination of cobalt motivated the study of the solubility of dimethylglyoxime in ammonium and alcohol solutions by the use of amperometric titration with the dropping mercury electrode. Two theoretical articles were published on the same question, in which mathematical models for describing the solubility of dimethylglyoxime in the mixed solvents were proposed.

The following article in 1946 was dedicated to the explanation of the special properties of the catalytic wave of hydrogen, where the quantitative theory of the phenomenon was developed. A mathematical model of the catalytic wave was obtained, which was in agreement with the experimental evident about the influence of different factors on its height and form. Apparently, this was the first mathematical model of the catalytic wave in polarographic analysis. Later Brdichka (1947) and S.G. Mayranovskiy (1957–1960) developed the theory of the catalytic currents of hydrogen, caused by organic bases, based on the idea of a reaction layer on the surface of the electrode.

The history of the third article is as follows. In 1944, Professor I.Ya. Postovsky (Ural Polytechnic Institute, chairman of the department of organic chemistry) turned to me with the proposal to study three groups of organic compounds obtained by him. These substances had different substituent groups attached to them. The purpose of the experiment was to obtain a correlation between the substituent a half-wave potential and the biological activity (drug-induced action). I willingly agreed to this proposal and appointed my colleague L.M. Reynus to obtain the polarograms of these substances: quinones, phenones and fuchsones. Apparently, this article was the first in USSR on the establishment of a correlation between the half-wave potentials (oxidation-reduction properties) for a number of organic substances containing different functional groups, and their biological activity.

The fourth article — also written together with L.M. Reynus — was dedicated to a polarographic study of solutions of niobium in nitric acid. The basic purpose of the work was to create the procedure for the polarographic determination of niobium (this procedure was necessary for other studies conducted in the institute). It was not realized since niobium compounds are hydrolyzed in dilute solutions of nitric acid and the wave height of niobium gradually decreases. Moreover, the wave of niobium could not be obtained using other dilute acids as supporting electrolytes. This work was subsequently developed further. It turned out that stable solutions of niobium with clearly defined waves are obtained in concentrated solutions of acids. E.I. Krylov (1954) proposed the use of 70% solution of sulfuric acid, and D.I. Kurbatov (1958) introduced concentrated solutions of pyrophosphoric acid (12–23 mole/litre). Then over a number of years [1960–1990] D.I. Kurbatov developed numerous procedures for the determination of different transition metals encountered in ferrous metallurgy, all of them measured using a background of concentrated solutions of different acids.'

Such articles formed the basis for Stromberg's doctoral thesis. Other articles from this time appeared in the journal *Plant Laboratory* concerned with the details of the polarographs built under his supervision and with the determination of copper and nickel in steel. The latter was carried out at the request of the central laboratory of the Ural Heavy Machine

Building Plant and was also reported by Stromberg at a conference in Kiev on physical chemistry methodology. He also relates other, compulsory, applied research undertaken in the late 1940s.[27]

> *'By 1950 the necessary material for my doctoral dissertation[28] was essentially completed. In the summer of 1949 for several months prior to testing of the first A-bomb, I was directed to "Chelyabinsk-40" town for three months, where the final preparations were being completed for making the materials for its production. It was necessary to install procedures for the polarographic determination of the purity of uranium and other elements in the materials for the bomb.'*

Chelyabinsk

In late 1945, soon after the United States used atomic warfare to destroy Nagasaki and Hiroshima in Japan, about 70,000 inmates from 12 labour camps were deployed along the banks of the River Techa on the eastern side of the Southern Urals to start the creation of what was to become a sprawling, secret, closed city to produce weapons-grade plutonium. The city would become known as Chelyabinsk-40; later it was called Chelabinsk-60, and from 1994 it was renamed as Ozyorsk. Some believe it now to be the most environmentally contaminated location on the planet.

Chelyabinsk-40 (Ozyorsk) is located about 15 km east of Kyshtym, about midway between the cities of Sverdlovsk (now Ekaterinberg) and Chelyabinsk. It occupies an area of around 200 km^2 between three lakes in the upper Techa River drainage basin where there are numerous interconnected lakes. Some 90km^2 of the town are occupied by the plutonium facility which ultimately utilized five uranium-graphite reactors (A, IR-A1, AV-1, AV-2 and AV-3) built in the years between 1948 and 1955. They were variously closed during the period between 1987 and 1990.

[27]A.G. Stromberg, '40 years of the Special Research Laboratory', TPU, 2004.

[28]This thesis was that for the degree of Doctor of Chemical Sciences, the equivalent of a DSc in the British university system.

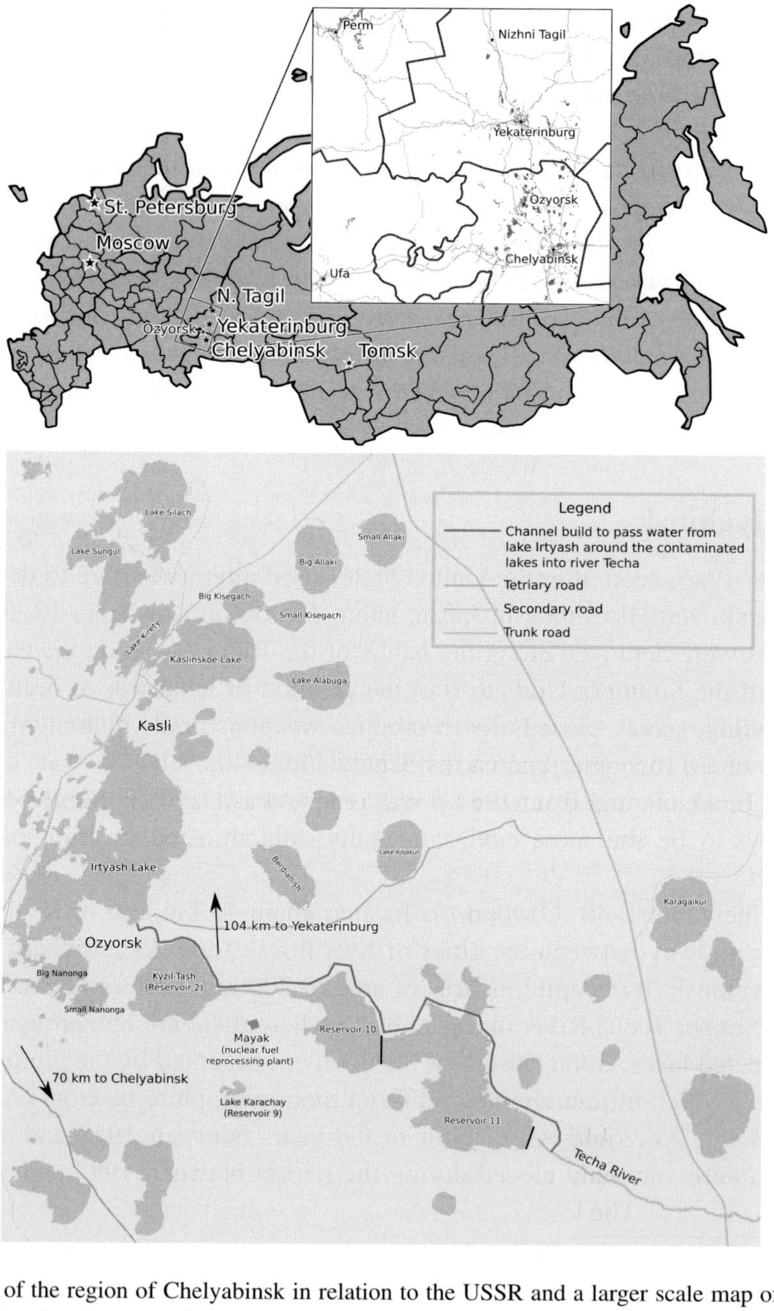

Maps of the region of Chelyabinsk in relation to the USSR and a larger scale map of the area showing the reservoirs and water channels built to contain the radioactive waste.

Other sites in the area were concerned with fissile material storage and processing. A reprocessing unit for irradiated fuels was built in December 1948 and the resulting high-level waste dumped straight into the adjacent Lake Karachai. The Tatysh plant, built in 1948–1949, about 20 km away from the reactor complex, produced plutonium and manufactured high-energy uranium weapons components. Today Ozyorsk is a town of approximately 90,000 inhabitants; the Mayak plant continues to process fissile materials and to make weapon components. Ozyorsk remains a closed city.

In the early years of operation, radioactive waste was dumped directly into the Techa River, which runs into the Tobol and Ob Rivers and ultimately into the Arctic Ocean. It has been estimated that some 2.75 million Curies of waste were disposed of in this way from 1956. Later, for a while, waste was diverted to Lake Karachai which accumulated an estimated 120 million curies of radioactivity; from the spring of 1968 this lake began to dry up, creating radioactive dust which was swept away by the wind showering an estimated 5 million curies of radiation on around half a million people. Between 1951 and 1954 a series of reservoirs were constructed along the Techa River, including Reservoirs 10 and 11 (see above map), and the river diverted around these reservoirs so as to retain some of the toxic waste already released.[29]

The uncontrolled release of radioactive waste into the Techa River system in the early years of operation together with primitive working practices saw thousands of workers — including alleged 'volunteers' from the camps who agreed to work with plutonium in exchange for a lesser sentence — and local populations exposed to dangerous levels of radioactivity leading to chronic radiation sickness, cancers, leukaemia and the relocation of thousands of local residents.

In September 1957 the cooling system of a containment unit malfunctioned leading to a major chemical explosion; Chelyabinsk, Tyumen and Sverdlovsk oblasts received an estimated 20 million curies of radiation affecting around a quarter of a million people and an area in excess of 20,000 km^2. The level of contamination was such that some areas today remain dangerous to human health.[29]

---

[29]http://www.globalsecurity.org/wmd/world/Russia/Chelyabinsk-65_nuc.htm

Armin Stromberg gave details of his secondment to Chelyabinsk in an interview in late 1997 to a family friend, Olga Skryabina, who recorded the recollections which translate as follows:

> *'Sometime around 1925 my stepfather Alexander Vasilyevich Vorobyev and I (then a 15-year-old boy) were invited to go fishing with his friend near the village of Kasli. He was an engineer at the local plant of castings, and he also invited us to visit the plant where he showed us the technology. Then we went fishing to Lake Kaslinskoe on two boats for the whole day from the morning to the evening.*
>
> *After the war, government selected this region for the production of nuclear weapons. In 1949 I [Armin Genrikhovich] was required to spend four months in Chelyabinsk-40. This was a new city and a plant for the production of nuclear raw material on one of the lakes in this region. At the end of 1949 the first nuclear bomb was detonated; it was prepared from this raw material [made in Chelyabinsk]. [And so what was] previously a natural paradise became a nuclear hell by the end of the 20th century. [In particular] a large river, the Techa, ran through there. The plant simply poured the contaminated waste into that river. All this was hidden from the community.'*

Olga Skryabina also provided the following summary of her conversation with Stromberg:

> '**Details of Stromberg's mission to Chelyabinsk-40.** *After the explosion of the nuclear bomb in Japan, the Soviet Union stimulated work in the late 1940s on the production of nuclear weapons. The bomb was created in the USSR in four years. The government did not stint on providing any and all resources needed to achieve this. Different physico-technical departments were created in technical educational institutes for training personnel for the atomic industry. A large number of plants were built for the production of nuclear raw material. Entire small towns appeared near the plants literally in one year. AGS*

worked in one of them in the South Urals for four months as one of the analysts for organizing the control of raw material for the A-bomb production. It was several months before the explosion of the first bomb. In Moscow there was a special analytical centre for the control of nuclear raw material named VIGAC headed by academician Ivan Pavlovich Vinogradov. The right was given to this academician to recruit any member of staff from any institute of the Soviet Union to take part in the control of nuclear raw material at any newly constructed plant. Thus Armin Genrikhovich was directed by Vinogradov's order of almost military status (without any personal agreement) to go and work in the plant at Chelyabinsk-40 (The latter was a totally absurd name, as it was located at a significant distance from the city of Chelyabinsk).

Precautions to ensure the strict secrecy of production. *[To illustrate the great secrecy surrounding the work], for example, Armin Genrikhovich had an entry document [to pass through] three checkpoints before he reached his laboratory. There were special signs on the passage document. On the first control point it was obligatory to fully undress, to leave all belongings in a special cabinet on a shelf, to pass into another room and [to ensure nothing had been secreted in the rectum] to squat naked several times with hands spread wide on the order of guard. Then people passed to another cabinet that provided state clothing, dressed again and went further along the corridor. On the way there were two additional controls, where everyone was required to show their documents. Each time the document was taken and [the holder's] face was compared closely with the photograph on the document. This entry document gave permission to pass only into the laboratory of electrochemical method of analysis. In contrast, staff working on the spectral method of analysis had a completely different document and had no chance to pass into another laboratory. People did not communicate with each other whatsoever, and did not know other's work at all.'*

Stromberg then moved on to discuss the speed at which the construction went on in and around the plant. He was able to estimate the rate of construction of highway:

> *'An excellent highway several kilometres in length appeared literally in several days in a place where there was previously nothing. In the process of building it large machines scurried continuously depositing crushed stone and asphalt. By the end of the day on returning from work there had appeared an excellent highway where in the morning there had been only empty space. All these jobs were done by prisoners who were brought from the numerous neighbouring camps. The importance of these works was indicated by the fact that the academician Vinogradov lived in the city, directly supervising all these tasks personally. Several times AGS talked with him, and was allocated specific tasks. A separate cottage was specially given exclusively for the use of Vinogradov, and only special personnel visited this.*
>
> *The rewards of this trip were very advantageous: Armin Genrikhovich received double payment (one salary for the post of the manager of the laboratory in the research institute and one for his work in Chelyabinsk-40). Moreover, the payments of the managers of laboratories in the academic institutes connected with the production of A-bomb were two times higher than the payment of a chairman of a department in any education institute. Money spent on the creation of the nuclear industry was expended without any thought.'*

This post-war period saw Stalin inspire Union-wide hostility to all things foreign, including scientific results and discoveries so that Russian scientists were seen exclusively as the inventors of all new ideas and technologies — an outlook sarcastically captured in the phrase 'Russia — the native land of the elephant'. As manager of the analytical laboratory Stromberg was required to make regular reports to those running his institute. The director was I.P. Bardeen, a metallurgist and academician. He, however, lived and worked almost entirely in Moscow and visited Sverdlovsk but once or twice a year. In reality, his deputy, N.V. Demenev fulfilled the day-to-day duties of running the institute. Stromberg relates one occasion in 1951 when he was required to report during a visit by Bardeen:[30]

> *'At the mentioned conference all managers of the laboratory had to make reports. At that time I was unaware of the political situation and prepared a report entitled "Successes in the development of polarography", in which references to foreign works occupied a fitting (and significant) place, and which showed our* [that of the USSR] *delay in this sphere. Fortunately, during the first day my turn to speak did not arise. But I understood from the reports of the managers of the other laboratories that the management of the institute would not welcome the content of my report, notably I.P. Bardeen. So within one evening (and part of the night) I completely reworked the report and excluded all references to foreign scientists. The report created a good impression and I.P. Bardeen noted the successes of the analytical laboratory in his closing speech.'*

Stromberg was nevertheless in a precarious position. Shortly he would be dismissed from his position of manager in his research institute.

Stromberg is Sacked — An Academic Life Begins

Notwithstanding the implicit recognition of Stromberg's skills as a top-rank analytical chemist in his being chosen by Vinogradov for the Chelyabinsk-40 project, in 1950 Stromberg was dismissed from his post at his academic institute for being a 'politically undesirable element', that is to say, in this case purely and simply for having German ancestry. Nevertheless he was able to move to the Ural State University (UrSU, Sverdlovsk City) for the next six years, first as an associate professor and then, in 1954, as a full professor. Here he lectured and taught a range of physical chemistry and electrochemistry to undergraduates whilst he himself was also studying hard, albeit not chemistry, since he was 'obliged to pass a three-year course of evening study at the Marxist-Leninist University (VMLU)'.

---

[30]Translated from A.G. Stromberg, *Russian Journal of Analytical Chemistry*, 48, (1993), 939.

Stromberg himself worked extensively on 'amalgam polarography' where metals pre-dissolved in mercury are oxidized and this was the substantive content of his thesis. He seems to have become interested in this topic around 1946 as a result of reading a report of the work of the American polarographer, Lingane, who has dissolved cadmium in mercury and then by polarography recorded the oxidation wave for the cadmium in an aqueous solution of potassium chloride. Stromberg, with his student A.I. Zelyanskaya reproduced the experiment but insightfully added cadmium chloride to the solution. A single polarographic wave was recorded with the current changing sign as the scan went increasing negative with the oxidation of the dissolved cadmium being seen at positive potentials and the reduction of the dissolved cadmium at relatively negative potentials. This suggested to Stromberg that the electrode kinetics of the cadmium/cadmium (II) couple were fast so that the wave was 'electrochemically reversible'. Stromberg and Zelyanskaya then carried out the analogous experiment using zinc and a solution of zinc (II) in aqueous ammonia. This time two distinct waves were seen; one due to the oxidation of zinc in the dropping amalgam, and the other resulting from the reduction of the dissolved zinc (II). They concluded that in this case the kinetics was much slower and that the process was 'electrochemically irreversible'. These are elegant and conclusive experiments. Despite the potential for exploring electrode kinetics in this way work was stopped on this topic for a few years, probably as a result of the analytical pressures in the research institute related to practical matters and, not least, supporting the A-bomb project.

In 1949, however, Stromberg decided that amalgam polarography would be the main theme of his DSc dissertation. Before doing so he consulted more experienced scientists, turning both to A.P. Vinogradov and to the Moscow-based academician A.N. Frumkin. He received opposing views. The former was consulted during Stromberg's four-month stay in Chelyabinsk:

> *'Vinogradov was strongly opposed to my intention. To some extent, from the point of view of analytical practice, he was right, since the idea of the electrochemical concentration of the metal in the form of amalgam on a stationary mercury electrode matured*

A.N. Frumkin, the father of Russian theoretical electrochemistry.

> *only much later (1957–1958). The question of how amalgam polarographic analysis might find practical usage for analytical purposes also, of course, occupied me, and I made some attempts to find such a use. In the work* [published in 1956] *I dissolved metallic indium in mercury (it dissolves very well) and then observed the anodic waves of admixtures using the amalgam dropping electrode. However, the procedure had obvious deficiencies for analytical practice.'*

Frumkin offered an entirely different perspective, from a theoretical rather than applied outlook. Without a doubt the greatest Russian electrochemist ever, Frumkin had a strong interest and background in fundamental and theoretical electrochemistry and immediately grasped the possibilities for the exploration of various problems in electrode kinetics such as the measurement of transfer coefficients and of exchange currents. He had been elected professor at the Lomonosov Moscow State University in 1930 where he was the head of the Electrochemistry Department that he had founded. His election as a full member of the Academy of Sciences of the USSR was in 1932.

Frumkin himself made many contributions to understanding fundamental electrochemistry, not least the introduction of the concept of the potential of zero charge and showed how the structure of the electrode-solution interface influenced the rate of electrode reactions. The 'Frumkin effect' describes how the concentration of dissolved electrolyte on the solution side of the interface changes the observed rate of an electrode process and this remains his most famous and lasting contribution along with the Frumkin isotherm which describes how molecules can interact directly with an electrode surface. Innovatively and realistically his model allowed for interaction between the molecules themselves as well as between them and the surface.

Following a conversation between Frumkin and Stromberg in Moscow in 1950 the former wrote a formal letter of support indicating a completely positive opinion about the latter's proposed direction of scientific research. This was crucial in resolving the formalities required for the defence of Stromberg's thesis, 'The theory and practice of polarography and amalgam polarography' in late 1951.

> *'Since my doctoral theses contained three clear distinct parts, the doctorate board recommended three different specialists as the official opponents: S.G. Mokrushin, head of the department of physical and colloid chemistry in Ural Polytechnic Institute (application of polarographic analysis for the resolution of physical chemistry problems); O.A. Esin, head of the department of the theory of metallurgical processes in UPI, a specialist in theoretical electrochemistry (theory of amalgam polarography); N.A. Tananaev, head of the department of analytical chemistry in UPI (application of polarography to analytical chemistry). N.A. Tananaev initially was doubtful about becoming an official opponent, but when I showed him the written opinion of A.N. Frumkin [his] doubts disappeared.'*

Six months after the award of the degree of DSc was confirmed by the relevant authorities in Moscow, Stromberg was awarded the title of full professor and immersed himself into his research and teaching.

> *'From that moment on, no one forced me to choose the direction of my studies, and during the next three years I obtained fundamental results on the study of the basis of, and new scientific applications for, amalgam polarography; although, generally the working conditions were not very favourable. Educational and administrative tasks took much time and effort (after moving to*

work in the university): mastering of new lecture courses (physical chemistry, theoretical electrochemistry, colloid chemistry, the theory of solutions) as well as practical and laboratory training. I didn't have any vacations in which to prepare the defence of my doctorate thesis. Furthermore, as soon as I had defended my doctoral thesis, the Rector required me to attend evening classes for three years on Marxism–Leninism. Many of my university colleagues managed to avoid this duty under different pretexts. But for me, being vulnerable as a result of my nationality, it was absolutely necessary to dutifully attend all the classes, to carry out the set homework, to be prepared for the seminars and to pass the examinations. It is noteworthy, however, that in those years (1952–1955) the evening classes were approached very rigorously by both the teachers and the students in contrast to the more routine approach that evolved in later years; the studies of Marxism–Leninism were time-consuming and exhausting.

In spite of these circumstances, these three years were probably the most productive in my life in the sense of theoretical significance and the novelty of our experimental results.'

Two principal results stand out from this work. First, pioneering work on the electrolysis of complex ions was conducted. Here a metal combined with various ligands, a complex, might undergo direct electrolysis to form the metal (dissolved in the mercury droplet), or it might lose all the ligands prior to electron transfer and undergo reaction as the simple hydrated cation, or in a third possibility some but not all of the ligands would be lost prior to electron transfer so that the electro-active species would be an intermediate with a structure which was neither that of the initial complex nor that of the simple hydrated cation. By studying systems such as that of zinc (II) coordinated with ammonia, Stromberg and his colleagues proved that this third possibility could take place; previously opinion had been split between the other two cases. They recognized that a spectrum of behaviour operated and that in any particular example any one of the three situations could represent the actual mechanism.

Second, Stromberg justified Frumkin's confidence that the area of amalgam polarography would be fruitful for the study of electrode kinetics by examining systems in which both anodic and cathodic waves for the same couple were present, as in the example of zinc studied by Stromberg and Zelyanskaya in their initial experiments using amalgam polarography. The measured difference in the half-wave potentials of the two waves allows one to evaluate the value of the exchange current (once the relevant theory had been developed), which is the fundamental measure of the rate of the electrode process.

Despite these successes, as Vinogradov had pointed out when asked in Chelyabinsk about the likely value of amalgam polarography, the method lacked at this time much authentic analytical value. This situation completely changed with the work of Barker and Kemula and the switch from the dropping mercury electrode to a (stationary) mercury drop (*vide supra)*. By holding the electrode potential at a negative value, metal ions present at very low level in a solution could be reduced to their metallic form and concentrated inside of the drop. This accumulation could, with a stationary drop, continue for tens or hundreds of seconds so as to build up a sufficient concentration of metals in the drop so that when the electrode potential was swept to more positive values the oxidation of the metals is apparent and a signal generated which can be used to quantify the level of ions present in the original solution. Anodic stripping voltammetry was to become a major analytical tool. In 'half a century with polarography' Stromberg reflects on why, in over a decade of working with amalgam polarography and searching for analytical uses for it, he and his co-workers missed the discovery.

> *'When, from reading the foreign scientific journals, I became aware of the principle of electrochemical accumulation (preconcentration) in amalgam polarography, I frequently asked myself: why in those years did I not hit upon the idea that would seem so simple and obvious? And I came to the conclusion that there were two reasons for this. One is the limitation of our thinking in connection with the use of the dropping electrode. The drop of amalgam, escaping from a capillary renews itself every two to four seconds whereas* [in anodic stripping analysis] *accumulation is conducted for hundreds of seconds. The replacement of the dropping amalgam electrode by a stationary mercury drop seemed*

to go against the very heart of the principles of polarographic analysis.

The second reason may be more important. It was the absence in the Soviet Union at that time of analytical equipment that possessed a signal recorder (plotter) in combination with a continuous linear change in the potential, [which we could have] applied to a stationary amalgam electrode after the accumulation of the determined metal in the form of an amalgam. The production of the Odessa photographic registering polarographs was discontinued after the war. All measurements were conducted with the use of the polarographs with reflecting mirror displays [the construction of which was discussed above].*'*

In 1956 Stromberg was offered the opportunity, by the rector of Tomsk Polytechnic University, A.V. Vorobiev, to move east to the city of Tomsk to work there as the Chairman of the Department of Physical and Colloid Chemistry. The desire to leave Ekaterinburg in the fullness of time, seemingly to escape having to live in such close proximity to his various family relatives, has been expressed in several of his letters from the GULAG in Chapter 3; whilst an interview conducted by the authors with Stromberg's sister, Marina, who lives in Ekaterinburg, suggested that the Stromberg family encountered problems with neighbours in the flat where they lived. For example, these neighbours constantly listened to their radio at top volume, which prevented Stromberg from being able to work at home. When he asked them to turn the volume down they explained that their speaker, which would be plugged into the communal radio system of their block of flats, had no volume control. Shortly afterwards Stromberg presented them with a new speaker with a fully functioning volume controller. The neighbours took offence and matters deteriorated further. The family were very happy to move to Tomsk.

5

The Tomsk School of Electroanalysis

Early Years In Tomsk: 1956–1962

Stromberg found Tomsk Polytechnic University a considerable change from the academic surroundings in Sverdlovsk. In his own words[1]:

> *'In 1956 I moved to the city of Tomsk. From that year until the present time I have worked at Tomsk Polytechnic University (TPU) in the Sub-Department of Physical and Colloidal Chemistry as the chairman of the sub-department (for 30 years), and since 1985 till the time of writing as a "professor-consultant[2]".*
>
> *In 1956 I started my work at the sub-department. There were eight people on the staff (I was the eighth to be appointed). Only one lecturer had a doctorate degree in chemical sciences (this was the former head of sub-department A.S. Naumova). There was no common scientific subject in the sub-department. The majority of the lecturers worked on topics provided by other sub-departments. The staff of the sub-department had only the vaguest notions of the area of polarographic analysis and, of course, there was no polarographic equipment. The teaching of physical and colloidal chemistry was conducted at a lower level in comparison with the*

[1]Translated from A.G. Stromberg, '40 years of the Special Research Laboratory', TPU, 2004.
[2]The term 'professor-consultant' is one employed by Stromberg; his role was more than that of an emeritus professor as understood in the West, retaining and exercising the right to act as the official supervisor of doctorate students.

Sub-Department of Physical and Colloidal Chemistry at Ural State University (UrSU). The entire programme — teaching and scientific research — had to be started afresh from scratch. I spent three years doing this. The number of students — the audience for the lectures on physical chemistry — across three departments was at first 250 chemists, 25 physico-chemical technicians and 50 geological surveyors. In 15 years the number of students in the Chemical Engineering Department grew by a factor of one and a half to two times and in the 1970s totalled 400–450 people.

My experience of the teaching of physical chemistry in UrSU (lectures, seminars and laboratory practical classes) was transferred to the Chemical Engineering Department of TPU. I taught a cycle of lectures on physical chemistry for the chairmen of the TPU sub-departments so as to improve uniformity in the teaching of chemical disciplines throughout the department, including establishing common names and designations of physical-chemical constants. To broaden the scientific knowledge of lecturers within my sub-department I read several cycles of lectures on electrochemistry and polarographic analysis.

Gradually I transferred lecturers from the scientific areas of other sub-departments, into the study of polarography. I also acquired polarographic equipment. During the five-year period (1956–1960) two lecturers of the sub-department defended doctoral theses, one under my full supervision (A.I. Kartushinskaya) and the other with my partial supervision (Kh.A. Lelchuk).

I made significant efforts to obtain more staff to carry out scientific work in the Sub-Department of Physical and Colloidal Chemistry. On my request the academician I.P. Alimarin wrote a letter to the rector of TPI,[3] *A.V. Vorobyev, on behalf of the Commission for Analytical Chemistry in which he recommended that Tomsk Polytechnic Institute develop studies in the area of the new highly sensitive electrochemical methods of analysis for the quality control of the purity of electronic materials. The letter*

[3]Named as Tomsk Polytechnic Institute between 1944–1991, later renamed Tomsk Polytechnic University, TPU.

made a strong impression on the rector, and he immediately gave me funding for four scientific researchers, seconding them from the Special Research Laboratory for the Study of Polymers. Soon afterwards the USSR Ministry of Education funded eight scientific workers so as to strengthen the "Scientific Research Institute for the Control of Electronic Engineering Materials". Immediately, the rector reclaimed the original four posts! Thus, within five years of my arriving in Tomsk I had eight scientific researchers for polarographic studies (in my sub-department).'

These 'new' highly sensitive electrochemical methods of analysis were in fact based on the stripping analysis with a stationary mercury drop first conceived by Barker and then realized by Kemula. It seems that Stromberg first learned of this major new advance from a discussion with Alimarin[4]:

'In 1958 I met I.P. Alimarin in Moscow State University, MSU in the Department of Analytical Chemistry. Ivan Pavlovich told me

Stromberg (centre front) and members of his sub-department in 1960.

[4]Translated from A.G. Stromberg, *Russian Journal of Analytical Chemistry*, 48, (1993), 939.

Sometime later, Rector Vorobyev was visiting this Ministry and heard about a decision of the government concerning the creation of six new special research laboratories in the education institutes, especially in electronic engineering. Immediately he submitted my request for the creation of such a laboratory at TPI, although my research was not directly concerned with the production of electronic devices. Nevertheless, the Ministry of Higher Education included TPI in the list of six educational institutes for the organization of such new laboratories.

After a while, Rector Vorobyev called me to a meeting of a group of chairmen of departments and told us his plan to create an additional (fourth) institute "in electronic engineering" [in TPI]. *At that time the department for radio electronics was part of TPI (although later it evolved into the TUSUR institute). My request for the creation of the laboratory for the development of new highly sensitive methods of analysis was the basis of the planned institute.*

Vorobyev's plans were immensely ambitious: the "Special Research Laboratory" was to consist of ten divisions with a total staff of 100 people. The request for equipment amounted to several million roubles. The construction of a new building was proposed. I was extremely concerned, since I anticipated that in this institute my laboratory would likely become a service unit, to satisfy the needs of the other laboratories. Rector Vorobyev asked me if I was in agreement with his plans. I screwed up my courage and said something along the following lines: "For seven years I was the manager of the laboratory of analytical chemistry in an academic institute (in Sverdlovsk) and I am sick and tired of making analyses for other laboratories. If the Rector intends to give this role to my laboratory in this new institute, then I ask not to be included in this project" (but I was shaking with fear since I knew that Vorobyev hated objections). Rector Vorobyev replied coldly: "No one intends you to serve other laboratories." I was relieved and said that in that case I agreed. Vorobyev sent the request for the new institute to Moscow.

Soon an answer came back from Moscow, saying that the Ministry had spent almost all the money given by the government for the creation of the other five special research laboratories, and therefore it could organize only a small laboratory in response to our request. So it allocated 320 thousand roubles[8] *for the equipment and payment for three units of staff.*

Vorobyev then took the opportunity to gain some benefit for himself. He proposed that I go to Moscow, along with one other colleague, and deal with the necessary papers for creating this special research laboratory. At the same time he gave me a list of extra new equipment for the Laboratory of Dielectrics and Semiconductors, in which he worked, to the tune of 150,000 roubles, and obliged me to include this list in the main list of equipment for my new laboratory.

By this time the total sum for equipment had been reduced by the Ministry down to 270,000 roubles, so that in effect my share was decreased to just only 120,000 roubles.

In December 1962 an order from the Ministry of Higher Education created the special research laboratory with the name "Special Scientific Research Laboratory of Physical Chemical Methods for Determining the Micro-impurities of Semiconductors and High-purity Materials" (in short: the Special Research Laboratory of Micro-impurities). As is evident, the name of the Special Research Laboratory of Micro-impurities does not correspond to the preliminary aim — the production of electronic instruments — for which the government created the six new laboratories. Later, in the middle of the 1970s, this circumstance nearly led to the closure of my special research laboratory after one inspection of its work by the Commission of the Ministry of Higher Education.'

The first successful experiment on stripping voltammetry in Tomsk was carried out by an assistant lecturer in Stromberg's sub-department, V.E. Gorodovykh, who had been recruited from Tomsk State University.[9]

'Certainly, a memorable moment in the history of the Special Research Laboratory of Micro-impurities was when we obtained the

---

[8]The equivalent of *c*.£700,000 today.

[9]Translated from A.G. Stromberg, '40 Years of the Special Research Laboratory', TPU, 2004.

first analytical signal from a micro-impurity and the publication of the first article on stripping voltammetry. In those days the method was known in the Soviet Union as "the method of amalgam polarographic analysis with pre-concentration". Our first article (and the first in the Soviet Union) was published in 1960 with two authors — V.E. Gorodovykh and myself — in the journal Zavod. Laboratory (1960. Vol. 26, pp. 46–48.). It was titled "The polarographic determination of 10^{-5} M of lead". In that same year in the Soviet press two additional articles using the SV method appeared: an article by two members of the Department of Analytical Chemistry of Moscow State University in the Journal of Analytical Chemistry and an article by a member of the academic institute GEOKHI in the journal Reports of the Academy of Sciences. The development of stripping voltammetry in the Soviet Union began with these three articles.'[10]

PhD student Vladimir Gorodovih in 1960, working at a polarograph called LP-55.

[10]The other two articles are: E.N. Vinogradova and G.V. Prokhorova, Zavodskaya Laboratoriya, 26, (1960), 41; S.I. Sinyakova and Yu-Wei Chen, *Reports of the USSR Academy of Sciences*, 131, (1960), 101.

The technique was to be used extensively over the coming three decades.

The Special Research Laboratory: 1962–1985

Stromberg was the scientific leader of the Special Research Laboratory of Micro-impurities from its founding in 1962 up until 1985 corresponding to his formal retirement. The role then passed to Professor A.A. Kaplin until 1989, after which the leadership was assumed by Professor Dr. Yu A. Karbainov when the Special Research Laboratory was subsumed into the Analytical Chemistry Department of Tomsk Polytechnic University. The laboratory throughout its lifetime strived to realize teaching and research of the highest standards. Stromberg has identified six principles underpinning his leadership and management of the Special Research Laboratory to ensure the 'harmonious combination of scientific research and teaching'. These were summarized in an article in the *Russian Journal of Analytical Chemistry* in 1991.[11]

> ***'Principle 1: The comprehensive development of a single method in analytical chemistry.*** *From the moment of its creation, the entire staff of the Special Research Laboratory were exclusively concerned with the development of stripping voltammetry. The theory of the method, the necessary equipment and practical applications were all realized. The range of elements of the periodic table amenable to determination by stripping voltammetry was enlarged and improvements in the sensitivity of their determination made. The narrowness of this focus had the benefit that the laboratory was at the leading edge of the development of stripping voltammetry throughout the period of its existence.'*

Whilst this benefit was clear, it might be argued that the evolution of the science was impeded. In particular, stripping voltammetry in its classical form could not and would not survive forever as a state-of-the-art analytical technique, and so this principle may have inadvertently sown the seeds that led to the Special Research Laboratory of Micro-impurities

---

[11]Translated from A Stromberg, *Russian Journal of Analytical Chemistry*, 46, (1991), 620.

A photograph of the staff of the Special Laboratory and the Sub-Department of Physical and Colloidal Chemistry in 1970. This photograph, with Stromberg featured in the centre front, was taken in front of the Chemical Technology Faculty, Building I, TPU.

ultimately being subsumed within the broader work of the Analytical Department of TPU. By the end of the lifetime of the laboratory in the 1980s, electrochemists elsewhere were variously developing the printed strip electrodes (screen-printed electrodes), now commonly used by diabetics the world over to assess their blood glucose levels, and also developing methods to harvest solar energy, and to make gas detectors such as those routinely used in smoke detectors in homes and offices. Stromberg's exclusive fixation with stripping voltammetry almost certainly prevented any possible diversification within the Tomsk electroanalytical sphere into areas where many of the principles and lessons learned by his team over the years would surely have generated potentially major scientific impact.

> ***'Principle 2: The combination of theory and practice.*** *The laboratory combined basic research with applications and, by means of contract research, put developed analytical procedures into the laboratories of industry and of scientific research institutes.'*

The role of stripping voltammetry, as with any branch of analytical chemistry, is to make reliable and useful measurements. Such a synergy as Stromberg describes was doubtless of academic benefit in that the posing of new analytical challenges would drive the scientific creativity of the unit to produce new instruments, procedures, techniques, etc. but also it represents the *raison d'être* of the entire scientific programme. Without an ultimate end user there is no point to analytical chemistry.

> ***'Principle 3: The combination of scientific research and teaching in the university.*** *The Special Research Laboratory was associated with three sub-departments (physical and colloidal chemistry, analytical chemistry, and electrochemical technology), which formed a single teaching and research unit concerned with voltammetry. This made it possible to engage teaching staff in the scientific research of the laboratory, and conversely, the laboratory staff in teaching. The general staff, together with the doctorate students and researchers, variously comprised between 50 and 70 people. Almost the entire staff worked in a single scientific direction and most of the teaching staff of the three sub-departments were former doctoral students of the Special Research Laboratory.'*

The potential synergy between teaching and research is one exploited in many systems of higher education where the use of top-class

> *with one of the [three sub-]departments. After completion of this period of postgraduate study (s)he is temporarily enrolled onto the laboratory staff to defend his or her dissertation, publish additional articles and conduct additional studies. After confirmation of the award of a scientific degree the new doctor leaves the Special Research Laboratory and must find themselves a place of work, according to the initial agreement* [see above]. *Until recently* [because of the shortage of skilled chemists] *in Siberia this was not a problem. Thus the basic output of the laboratory's activities is described in 87*[12] *PhD theses and three DSc theses, defended over the 27 years that the laboratory existed, and in the scientific articles published during the process of the preparation of these theses.'*

The structure of the doctoral programme described follows a structure commonly used in many Western equivalents with a significant taught course element in the early years before focusing on original research. The Engineering and Science Research Council in the UK has recently introduced the doctoral training centre scheme (DTC) where centres focus broadly on a specialized area (such as organic synthesis or biological chemistry). The doctoral candidates undertake formal tuition across the breadth of the specialization over the first year of their study, given by a range of experts, and the selection of their doctoral research topic is deferred until after this is largely completed. There are significant similarities with the Stromberg Special Research Laboratory approach.

The appendix at the end of this chapter lists all of the doctoral students from the Special Research Laboratory of Micro-impurities. Interestingly, no less than 66 of those named are female! It seems that male chemistry undergraduates had greater opportunity for employment in local industry where the salaries were higher, but which was male dominated since the work involved the running of heavy chemical plants. Those graduating with doctorates, typically around three per year, moved into other

[12]The number of 87 theses was that at the time of writing; the total number ultimately was higher as shown in the appendix given at the end of this chapter.

Table 5.2: The initial destinations of the first 87 doctoral students of the Special Research Laboratory of Micro-impurities.

Destination	No. of Students
Tomsk	54
Western Siberia (excluding Tomsk)	14
Eastern Siberia and the Far East	6
The Republics of Central Asia	3
European Russia	3
The Ukrainian SSR	7

scientific or educational institutes in Siberia. The initial destinations of the first 87 doctorates are given in the table above.

As is evident, some 74 former postgraduate students stayed to work in Siberia; only 13 people moved to work in other regions. Of the 87 former postgraduate students, 67 moved to work in educational institutes and universities (typically at a rank equivalent to associate professor), 18 went to applied research institutes and 2 went to academic research institutes. Finally it is worth pointing out that the figure of 87 researchers receiving their doctorates represents a 100% pass rate! There were no failures in the doctoral examinations for the simple reason that Stromberg would not let any student submit a thesis until he considered it effectively perfect; in practice this meant that he had to deftly shuffle his resources so that the students remained employed in the special laboratory until the necessary level of scientific output was accomplished, sometimes they were employed there for as long as nine years.

Of the three DSc graduates from the Special Research Laboratory up until 1985, Yu Karbainov, A.A. Kaplin and M.S. Zakharov, the first two remained in Tomsk (see above) whilst the latter moved in 1968 to work in the Tyumen Industrial Institute (which became Tyumen University in 1980) where he created a new laboratory, also specializing in the area of stripping voltammetry, which published around 120 research papers in its first 22 years of existence. Zakharov's DSc thesis was concerned with anodic stripping voltammetry applied to 'highly purified materials', those of Karbainov and Kaplin with applications to 'metal complexes and extractions' and 'crystals and thin films', respectively.

Armin Stromberg (left) and Anatole Kaplin discussing a polarogram in 1960.

'Principle 6: All studies must be regularly discussed and reviewed. *Scientific seminars and review meetings are the basic collective way of organizing work and disseminating information about all studies in the Special Research Laboratory and in the groups within that laboratory. Special attention is paid and great significance attached to seminars and meetings. Review meetings are called once every three months in which the leaders of the groups give talks, summarizing the groups work during the preceding period and presenting plans for the coming months. All members of the Special Research Laboratory make daily entries into their personal notebooks which also contain the quarterly plans and reports. Scientific seminars are held typically once every fortnight. Annual reports from the graduate students and colleagues are made at these seminars, with nominated 'opponents' to lead the discussion. In addition, there were reports about the content of articles prepared for publication with 'internal' reviewers selected from the more highly qualified scientists and teaching staff of the university giving their feedback. Overviews of particular problems in stripping voltammetry, information about recently published papers, and new abstracts appearing in the Russian Journal of Chemical Abstracts were regularly presented along with courses*

Table 5.3: The distribution of research papers amongst different scientific areas in the Special Research Laboratory of Micro-impurities.

Scientific Area	**Number of Publications**
Theoretical research	156
Improvement of equipment	59
New methods of research and analysis	92
Studies in physical chemistry	144
Conditions of chemical analysis	137
Development of stripping voltammetric procedures and protocols	129
In total	717

of lectures of various aspects of electroanalytical chemistry and more general issues of analytical chemistry.'

Principle six describes the way in which a large, modern research group would typically operate. However, the formal way in which this was structured predates such 'good practice' in Western Europe probably by one or more decades. Over the lifetime of the Special Research Laboratory it produced around one half of the total publications from the USSR in the area of stripping voltammetry. The distribution of the papers amongst the different scientific areas is in Table 5.3.

Life as a Graduate Student in the Special Research Laboratory

Several former students have described their experiences whilst studying in the Special Research Laboratory. These recollections were solicited for a celebration of the fortieth anniversary of the founding of the Special Research Laboratory of Micro-impurities. Accordingly, they are uncritically congratulatory in tone, but nevertheless convey the spirit in which the work was undertaken in the laboratory. The first is from Inessa P. Mamontova who obtained her degree in the Chemistry Department of Tomsk State University between 1963 and 1968 before joining the Special Research Laboratory of Micro-impurities. Subsequently, she changed scientific fields and worked in the Biochemistry Department of the Siberian State Medical University in Tomsk.

> *'Many years have passed since I was in the Special Research Laboratory, but I have always considered the people who work there as close colleagues. The friendly relationships and the wish to help one another surprised me from the day I joined the laboratory. I noted the clear, effective organization of the work of the staff. The system of signing in and signing out (arrival-departure) caused some irritations but also established a working discipline. Each colleague had a personal notebook in which their scientific* [group] *leader wrote a job schedule for a week or for a month, after which he required a report. Always some analysis was made following the report: if some experiment was not successful, then together they attempted to find new approaches to the solution of the problem. In this way a future scientist was formed from a girl without any prior research experience.*
>
> *Regularly, once a month, a scientific seminar of the laboratory was held. Armin Stromberg very skilfully divided the spheres of activity of A.A.Kaplin and B.F. Nazarov, and even the themes of the PhD theses were selected taking into account the specific talents of colleagues: some wanted more experimental work, and some — theoretical. Many significant problems were solved rapidly and collaboratively by a large group of colleagues working as a team. There was complete openness and each member knew what everyone else was doing. During the seminar it was possible to offer advice to a colleague or to criticize his or her work. Armin Genrikhovich encouraged this, but did not make any authoritative interventions during the discussions; only at the end did he voice his own opinion. Now, after working for 24 years in the medical school, I can appreciate the significance of our scientific seminars, which made it possible to solve complex problems together and to have a vision of the future. In medical science "corporate" behaviour is very widespread; everyone develops their ideas in secret and in fear that the material might be stolen. Probably science in medicine could be developed at a more rapid tempo and*

many more problems solved with a greater openness and an integrated approach.

We were trained to appreciate and use the current literature, to visit the library regularly, and to look through the Russian Journal of Chemical Abstracts and the current periodicals. Every member of staff periodically made reports on the various different areas of modern scientific investigations covered in the journals. The laboratory had its own library containing all the theses defended in the laboratory. This was very convenient.

People with dexterous fingers worked in the laboratory. The electrodes made in the laboratory not only performed excellently, but also were beautiful and elegantly constructed.

Even after we left the laboratory, the management was not indifferent to our fate. Armin Genrikhovich and Anatole Alexandrovich [Kaplin] *were always interested in our work and willingly offered advice. All of this makes it possible to consider the Special Research Laboratory a home from home. It is always a pleasure to visit and to know that they are always glad to see me.'*

A second reminiscence is that of Valeri A. Kolpakov who, after leaving the Special Research Laboratory of Micro-impurities, lectured at Tomsk Polytechnic University in physical chemistry as well as teaching the design of chemical plants and the 'technology of new materials'. Before joining the Special Research Laboratory in 1967 he was a student in the physico-technical department of TPI interested in the chemistry of radioactive elements. He left this position in 1966 and briefly joined an industrial (ship-building) plant some distance from Tomsk before joining Stromberg's laboratory.

'I joined the Special Research Laboratory in 1968 by virtue of my acquaintance with A.A. Kaplin, Yu.A.Ivanov and V. KIvanov: we shared a single room for a while in the hostel during my studies at the institute.

Lasting friendships were formed in the laboratory. We were young and unburdened by any serious difficulties; life was stable,

'It was 1967 when I met Armin Genrikhovich Stromberg for the first time. He was the chairman of the examination board for the defence of student diploma theses in physical chemistry in the Chemistry Department of the Tomsk State University. We [candidates] *were pleasantly surprised at the benevolence shown to us by Armin Genrikhovich; we did not expect this, having heard beforehand about this distinguished scientist, and about his scientific rigour.*

In February 1969, after various unsuccessful attempts to find a job as a scientific researcher, I had a long conversation with Armin Stromberg; afterwards I was accepted as a senior technician in the Special Research Laboratory of Micro-impurities.

Having had work experience in production at the Siberian chemical enterprise in Tomsk-7 and in the Branch No 1 of Tomsk Polytechnic Institute, I moved into a completely different atmosphere in terms of the relations between the co-workers in the laboratory. [...] *I understood that I arrived in a true scientific school where there was a scientific rhythm to life ... always with the expectation of the development of scientific knowledge. This was the Scientific School created by Armin Genrikhovich — the "School of A.G. Stromberg".*

It is difficult to describe the scale of effort, which was made by Armin Genrikhovich: the sequence of his actions in the organization of the Special Research Laboratory, in the creation of scientific collaborations, in the attraction of the teaching staff of the department to scientific work on the theme of the Special Research Laboratory, on the maintaining of close contact with the administrative aspects of the department and the host institute, the Academy of Sciences of the USSR and other institutes. This was the enormous organizational, scientific and training work of Professor A.G. Stromberg, which was within the power only of a distinguished scientist and a most excellent organizer and teacher.

All of A.G. Stromberg's studies were guided by single common idea, which created a school and which makes it possible to assert that his name will remain in the field of science forever.

The uniqueness of the Special Research Laboratory of Micro-impurities should be noted both in terms of its structure and its operating principles. These are as follows:

- *Development of the theory of stripping voltammetry;*
- *Study of physical-chemical effects with the help of stripping voltammetry and the practical application of SV in chemical analysis;*
- *The development of electronic equipment for stripping voltammetry.*

Such an integrated approach allowed Armin Genrikhovich to create sub-groups in various directions as indicated above. In due course the leaders of these sub-groups came from the postgraduate students of Armin Genrikhovich: A.A. Kaplin, B.F. Nazarov and Yu.A. Ivanov.

Another virtue of Armin Genrikhovich was the creation of a unique atmosphere in the Special Research Laboratory. As in any such large, active organization scientific disputes and conflicts of interests could occur, which unresolved might fester. On these occasions Armin Genrikhovich would contribute, for example, in our scientific seminars, from a non-partisan point of view reflectively and perhaps making analogies about the role of personality in history so as to defuse the situation with a minimum of confrontation. This [skill] *contributed much to the creation of a productive climate both in the Special Research Laboratory and at the department of physical chemistry.*

The results of the Special Research Laboratory under the management of Professor A.G. Stromberg testify to its success.

- *The staff of the Special Research Laboratory and department published more than 700 scientific publications;*
- *Tens of analytical procedures were developed for different analytes and used throughout the nation;*
- *World-class analytical equipment for stripping voltammetry was created;*

— *More than 100 highly skilled specialists were trained;*
— *A book, Physical Chemistry, written for university chemistry students and a book of problems in the chemical thermodynamics were published.*

The atmosphere in the Special Research Laboratory created by Armin Stromberg resulted in doctoral theses being realized within a period of 5 years, and of 3 years for [training university graduates for industry with the latter's financial support]. *Highly skilled scientific workers graduated from the Special Research Laboratory, with a unique schooling, who could successfully work both in scientific research or educational institutes.*

It should be noted that A.G. Stromberg's Scientific School had important off-shoots in the cities of Tyumen and Sverdlovsk. These were led by two students of Armin Genrikhovich — M.S. Zakharov and Kh.Z. Braynina.

The ideas and the scientific directions developed in the Special Research Laboratory were primarily not a continuation or development of the ideas of other scientists, but they belong to A.G. Stromberg. Therefore it can be said that in the field of electroanalytical chemistry there is a "School of A.G. Stromberg".'

The textbook on physical chemistry alluded to by Zheltonozhko was written in collaboration with D.P. Semchenko. The first edition appeared in 1974, running to seven editions, testifying to the 'classic' nature of the text. Indeed it was widely adopted — especially after excellent reviews of the third edition published in 1999 — and has survived as a useful teaching text until the present day.[14] The book is entirely predictable in structure with a division into quantum chemistry, chemical thermodynamics, chemical kinetics and catalysis, but is much more rigorous in the treatment of each topic and has a higher theoretical content than any of its contemporaries. The text was based on the course of lectures delivered by

[14]A copy of which occupies a space on one of the shelves in my office, and, although written in Russian, has been frequently referred to for handy solutions to physical chemistry problems even to this day — GGW.

Stromberg at TPI, but despite the depth and quality of the work it proved difficult to get the first edition published, probably due to a somewhat dismissive attitude of scientists based in Moscow to a text originating from a provincial institute of higher education. Yet the book has passed the test of time with a seventh edition appearing in 2009 and remains one of the best sellers on the list of the publisher (Vysshaya Shkola of Moscow, literally 'Higher Education'). The following is Stromberg's account of his struggles to have the book published.

Stromberg and Semchenko: *Physical Chemistry*

> *'The history of the publication of this textbook of physical chemistry is connected with a number of interesting incidents, which at times seem like fiction. In October 1969 I participated in a commission in Ivanovo town, created by the Ministry of Higher Education of the USSR for the development of a new programme of physical chemistry for the students specializing in chemical-engineering.*
>
> *It turned out that the head of the Department of Physical and Colloidal Chemistry of Novocherkassk Polytechnic Institute (**NPI**, Novocherkassk city) D.P. Semchenko had a completed manuscript on physical chemistry, which, in his words, required only "the smallest modernization", and that its modernization could be best accomplished by adopting some additional co-authors. The head of the Physico-chemical Department of Kuybyshev Polytechnic Institute (KPI, Kuybyshev city), Yartsev, (I have forgotten his name and patronymic) explained that the publishing house Vysshaya Shkola planned to publish a textbook on physical chemistry to be printed in the following year, but that there was, as yet, no author.*[15]
>
> *Semchenko and Yartsev agreed to jointly produce the new textbook using the manuscript of Semchenko as the basis, but requested me to be the editor saying that the addition of "several innovations to the already completed manuscript would be*

[15]In the USSR publications were planned even if there was no book in preparation!

required". I agreed, without suspecting what trouble these "several innovations" would become!

The work of three co-authors to modernize Semchenko's manuscript began in three different cities. Different parts of the manuscript were sent successively from Novocherkassk (from Semchenko) to Kuybyshev (Yartsev), and then on to Tomsk. Towards the end of review of the manuscript I realized first that Yartsev was adding nothing, but was only forwarding me the manuscripts without any changes; second, the manuscript of Semchenko was antiquated and was considerably worse than my lecture notes [that I was using to teach at TPI]*; I decided that to continue this task was pointless. I sent a letter of ultimatum to my co-authors telling them that either they should agree to use my lecture notes as the basis for the book and write the text according to my plan, or else I would leave the triumvirate. They agreed.*

So a new period in the writing of the book began. I set my colleagues tasks and gave them detailed directions to the sources they should use and described precisely the content of every chapter on a paragraph by paragraph basis (in accordance with my lecture notes). Again Yartsev did nothing, making various excuses — illness, travel, etc. and neglected to carry out any of the set tasks. So I wrote the necessary parts myself. Finally the manuscript was finished and sent to the publisher. In the end almost all of the manuscript had been written by me.

The editorial staff sent the manuscript to two reviewers (without any consultation with me). After half a year the reviews were completed and an editor called me on the telephone. The review of the "individual expert" (Professor L.A. Nikolayev, head of the department of chemistry of a military academy) was positive with a small number of comments. The review of the Department of Physical Chemistry of Moscow State University — actually from several sub-departments of the chemistry faculty — criticizing different sections of the textbook and signed by ten reviewers was crushing: "The manuscript is not fit for publication, and even after extensive revision will not be fit". It seemed then that the project was doomed and that the text would never be published.

However, the editor, perhaps not taking the second review at face value, allowed me to make a response to both reports.

After reading the second review (which was 40 pages long), I realized that the reason for the crushing report was primarily the chapter on chemical thermodynamics which, as I later discovered, had been reviewed by E.N. Eremin, a professor of the Moscow State University's Department of Chemistry. The criticisms of the other chapters of the book were trivial and did not correspond to the final damning conclusion. I accordingly wrote this in my response and awaited the editorial decision. The editor (in consultation with Professor L.A. Nikolayev) made a favourable decision: they would send the manuscript for additional review to the Department of Physical and Colloidal Chemistry of Leningrad Technological Institute (LTI) (Leningrad city), noting that Moscow State University was not a Technical Institute of Higher Education. This department at that time was managed by Professor Mashovets.

I rang him, and he invited me to visit him in Leningrad. He ordered different specialists in his department to examine various parts of the manuscript. Then each of them had a discussion with me. Over a period of three days the Leningrad review was written.

After receiving this additional review, the head of the editorial board for chemistry at Vysshaya Shkola made the decision to publish the textbook and we signed a contract.

[Since he had contributed nothing] *Yartsev should have been excluded from the contracted authors. However, legally this was only possible if Yartsev himself wrote to the publisher asking to be excluded. This seemed unlikely but a happy coincidence resolved the problem.*

I was in Moscow and met with the head of the editorial board of the publisher, and explained about the "Yartsev problem". She said that Yartsev was currently in Moscow, she had very recently met with him, he had presented her with flowers (!), and had just departed to a conference only a quarter of a kilometre from where we were. On her advice, I immediately left for the polytechnic museum and found Yartsev there. Soon after the conference session ended I told him that the head of the editorial board had

Stromberg working on his textbook in 1970 on the bank of the Ob River, whilst covering his head from mosquitoes.

asked to see both of us. Yartsev agreed (to my surprise), and he wrote the required statement in the presence of the editor.

Several months later an editor called me to Moscow to discuss the book. When I arrived she suddenly announced that she had looked again at the book and had concluded that it was not publishable and that all work on its production had ceased. I was bewildered until Professor L.A. Nikolayev explained that Professor Eremin had contacted the editorial board to suggest that if Stromberg's textbook were published then the Department of Physical Chemistry at Moscow State University might publish a damning review which would discredit the editorial board.

I realized that the only solution was to approach the head of the Department of Physical Chemistry of MSU, the academician Ya.I. Gerasimov, and ask him to confirm that I had considered all of the corrections suggested by his department, and that the rewritten version of manuscript, in his opinion, was publishable.

Gerasimov agreed to review the manuscript and I gave a copy to him.

Four months later I received a telegram from the publishing house: "Impression is mixed, please visit." I understood: the final key moment had arrived. If the new opinion was negative then the book would not be printed. I called Semchenko in Novocherkassk (Semchenko had considerable public standing — he was a member of the communist party, and formerly a rector of NPI), and we went together to see Gerasimov. Yakov Ivanovich Gerasimov proved to be an honest person. Excluding Eremin, he said: "I have examined the manuscript, and found it satisfies all the requirements." Next, he wrote a brief 20-line review. These lines solved the fate of the textbook.

I gave the review to the editor, the manuscript was "beautified"… everything went smoothly, and the textbook was finally published in 1974.

Let the readers judge for themselves, but it seems to me that the story is very illuminating, and it makes it possible to understand why textbooks written by non-Muscovites are so rare.

The second edition of the textbook came out 14 years after the first, in 1988. I wrote it myself, since Semchenko was seriously ill, and he died a year later. This edition was substantially rewritten. Its publication was made without any complications. I selected the reviewers myself, at the request of editorial board. To obtain a review from a department, I went to Dnepropetrovsk, to the Department of Physical and Colloidal Chemistry of Dnepropetrovsk Chemico-technical Institute, managed by Professor M.A. Loshkarev, whom I knew when he was the instructor of the Ural Polytechnic Institute in Ekaterinburg. Over a period of three days I talked with instructors in his department, who made observations on different parts of the textbook. I gathered these observations and, at the request of the head of department, wrote a review by myself. The review was discussed at a session of department, and then given to me.

In 1999 I wrote an additional chapter (including the theory of fractals) for the third edition). The fourth (2001) and fifth (2003) editions were published subsequently.'

In 1999 Vysshaya Shkola ceased to publish any alternative texts in physical chemistry or to commission new books in the area which might rival Stromberg and Semchenko.

> *'Now the textbook is adopted as the only textbook on physical chemistry for all chemical specialities in institutes of higher education in Russia.* [...]
>
> [After the first edition was published] *I asked some friendly associates among the original "ten" who had signed the first damning review (in particular from the Department of Theoretical Electrochemistry at Moscow State University), why they had signed it. They were confused, but explained that, on the assertion of E.N. Eremin, the section "chemical thermodynamics", which they did not see, allegedly, was terribly written, not suitable at all for publication, and they had simply believed him.'*

Stromberg's Reflections on Running His Laboratory and on His Relationships with the Scientific Community

As the saga of the publication of *Physical Chemistry* suggests, relationships between academics are rarely straightforward! In 2003, Stromberg wrote the following piece in two parts reflecting first on his lifestyle and secondly, presented here in a somewhat abridged form, on his relationship with the scientific community and other scientists. His observations make very good advice for any newly appointed or even experienced scientists!

> *'There are two types of people in their relation to science: (a) those who are actively and authentically scientific, searching for truth in science, and who understand the difference between real science and solecism; (b) those who do not distinguish between science and solecism or pseudo-science, and who see science as a cover or means for advancing themselves and their careers. I would gladly talk with the first, but break away from any relationship with the second, most often without explanation,*

but only very occasionally after making explicit criticism of them.

I have experienced two categories of people in terms of their personal relationships to me: (a) those who have treated me well and behaved decently; and (b) those who behaved somewhat cowardly in their relations to me, committing ill deeds, and behaving incorrectly. My attitude to them is as above.

I do not believe in relying on the spoken word. Always remember the parable about the three stages through which a new idea passes: 1. "It cannot be true!"; 2. "This idea is not so crazy!"; 3. "Everyone knows it is true!" Therefore, when I have a new idea, before proceeding with its development or experimental verification jointly with another scientist, I conclude a written agreement with the colleague, which sets out the essence of this idea, describes its novelty and its further stages of development or experimental verification, and notes who proposed the different parts of the idea and when. The agreement is then signed by both contracting parties (the generator of the idea and the developer or developers). Experience of life has convinced me that without this precaution, as a rule, misunderstandings often occur in accordance with the parable described above.

I use the same written planning in other situations (in the team), in the form of memoranda, or written report, etc., the important point being to always have a written statement.

In the event of a conflict with a member of staff or a colleague I never shout at them, nor try to antagonize them. I try not to show that I have noticed their actions against me. But afterwards I terminate any further communication with them (or any correspondence in the case of a distant person). They simply cease to exist for me. Life in Soviet times in the face of the Great Terror taught me such a "conformist" approach; false denunciation could always be expected from an aggressive, offended man, resulting in arrest and imprisonment in a camp.

In case of a deadlock in relations with a member of staff I try not to be volatile, whilst giving the impression that nothing has

Table 5.4: List of USSR enterprises and scientific research institutes collaborating with Stromberg's Special Research Laboratory.[16]

Name of an Organization	Year	Type of Work	PhD Student Responsible
Novosibirsk plant of Sn production, Novosibirsk	1962	Amalgam ASV detection of different contaminants (Cu, Pb, Zn, Bi and others) in purified Sn and In	L.F. Zaichko
The Central Laboratory of the Uzbekistan Ministry of Geology, Tashkent	1967	Ag and Hg in water; methods of deep deionization of water for analytical purposes Sensitivity: 10^{-7}–10^{-6} g/L	N.A. Kolpakova
Shymkent Lead Plant, Shymkent, Southern Kazakhstan	1967–1971	Determination of 10^{-8}–10^{-6}% Cd, In and Zn in purified Pb; determination of 10^{-7}–10^{-6}% Cu (1967), Sb, Bi (1970), Pb, Sn (1971) in purified Cd	V.E. Katuhin
Slavgorodsky Chemical Plant, Yarovoye, Altai region, Russia	1967	Unspecified in the contracts between the collaborators	Group of members of staff headed by Yu.L. Lelchuk and Yu.A. Karbainov
Sievierodonetsk Chemical Plant, Sievierodonetsk, the Eastern Ukraine	1968	Determination of 10^{-10}–10^{-9}% Cu, Pb, Cd, Zn in HNO_3	V.I. Kuleshov
Scientific Research Institute "Pulsar", Moscow (the central institute on production of transistors in USSR)	1968	Determination. of 10^{-6}–10^{-5}% Cu, Pb, Zn, Cd in films of AsGa	A.A. Kaplin
State Institute of Nitric Industry, Moscow	1968	Cu, Cd, Pb, Zn in HNO_3 and highly purified water (10–10%)	V.Z. Bashkatov

(*Continued*)

[16]Taken from thesis of former postgaduates and DSc.

Table 5.4: (*Continued*)

Name of an Organization	Year	Type of Work	PhD Student Responsible
Scientific Research Institute on Molecular Electronics, Moscow	1971	Detn. of 10^{-8}–10^{-7} % of Cu, Bi, Sb, St, Cd, Pb, In, Te, Zn, Mn in water to be used for the production of electric circuits	N.A. Pokrovskaya
Bashkir Scientific-Research Institute on Oil Processing, Ufa	1971	10^{-6} of V, Ni, Mn, Al, Ti, Cu, Pb, Si, Bi, and Co in coal, tar and gas oil	
Altai Scientific Research Institute of Chemical Technologies, Biysk	1972	Purity of explosives	A.A. Zheltonozhko
Novosibirsk Plant of Precision Engineering, Novosibirsk	1972	Development of Trilon+pyrophosphate electrolyte for Cd electroplating	N.I. Vahmantseva
Shymkent Lead Plant, Shymkent, southern Kazakhstan	1972	Determination of 10^{-8}–10^{-7}% Tl, Ni, Co in pure Cd	V.E. Katukhin
Scientific Research Institute on Molecular Electronics, Moscow	1972	Determination of 10^{-5} % Na and K in technological media (deionized water, HNO_3, HCl and other etching solutions); the polarograph model LP-60 was used in the introduction of the method	V.K. Ivanov, R.F. Zarubina
Unnamed plant "Postbox g-4147", Moscow (production of lasers)	1972	Amalgam-polarographic determination; Of 10^{-4}–10^{-7}% of Cu, Pb, Ni, Cd in salts and crystals of KH_2PO_4, KD_2PO_4, CsH_2AsO_4, CsD_2AsO_4, $LiIO_3$	N.A. Podkorytova, N.T. Rud

(*Continued*)

Table 5.4: (*Continued*)

Name of an Organization	**Year**	**Type of Work**	**PhD Student Responsible**
Unnamed plant "Postbox g-4147", Moscow (production of lasers)	1972	Determination of different contaminants in pure salts of CO_3^{2-}, NbO_3^-, TaO_3^- + alkali and alkaline; earth metals	Z.S. Mikhaylova, N.A. Podkorytova
Unnamed plant "Postbox g-4147", Moscow (production of lasers)	1972	Determination of 10^{-5}–10^{-6}% of Cd, Pb, Cu, Bi, Zn in SrF_2 for laser production	
Scientific Research Institute of Communication Devices, Omsk	1972	10^{-7}–10^{-5}% of different metals in evaporated films	N.P. Pikula
Institute of General and Inorganic Chemistry, Moscow	1974	Determination of 10^{-8}–10^{-7}% Li in crystals of Cd_3As_2 (thin layer)	N.K. Djabarova
Scientific Research Institute on Molecular Electronics, Moscow	1974	Detn. of 10^{-8} % Fe in deionized water, in HF, and in Si_3N_4 to be used at Different stages of the production of electric circuits	R.F. Zarubina, Z.S. Mikhaylova
Scientific Research Institute on Molecular Electronics, Moscow	1974	Determination of 10^{-5}–10^{-6} % of Cd, Pb, Cu in $CH_3C(OH)[P(O)(OH)_2]_2$	R.F. Zarubina
Scientific Research Institute on Molecular Electronics, Moscow	1975	Method of separate determination of Na and K in SiO_2 films (layers)	R.F. Zarubina

(*Continued*)

Table 5.4: (*Continued*)

Name of an Organization	Year	Type of Work	PhD Student Responsible
Scientific Research Institute "Pulsar", Moscow (the central institute on production of transistors in USSR)	1975	Method of separate determination of 10^{-6}–10^{-8}% Na, K, and Li	R.F. Zarubina
Scientific Research Institute "Pulsar", Moscow (the central institute on production of transistors in USSR)	1975	Determination of 10^{-7}–10^{-8} % of As, In in deionized water, HNO_3, and SiO_2	V.E. Morozova
Scientific Research Institute "Pulsar", Moscow (the central institute on production of transistors in USSR)	1975	Determination of Fe in SiO_2 films (layers)	L.F. Zaichko, R.F. Zarubina
Scientific Research Institute of Semiconductors, Moscow	1977	Determination of different contaminants in semiconductors	*V.E. Katuhin, *Z.S. Mihaylova, E.O. Portnyagina
"Plant of pure metals", Svetlovodsk	1977	Determination of Mn in pure Pb; Hg, As in pure Pb; Na+K in pure Pb, Hg and Cd	V.E. Katuhin, L.S. Anisimova
"Plant of pure metals", Svetlovodsk, Central Ukraine	1978	Purity of Ga: 10^{-6}–10^{-5} % of Zn, Mn, Te; Purity of TeP (Te phosphide): 10^{-6}–10^{-5}% of Zn, Ag, Cu, Bi, Te	*V.E. Katuhin
Scientific-production enterprise "Vostok", Novosibirsk	1978	ASV* determination of different contaminants in SiO_2 films: Zn, Pb, Cu, Cd, Bi, Sb, Au, Ag	V.N. Polyakova

(*Continued*)

Table 5.4: (*Continued*)

Name of an Organization	Year	Type of Work	PhD Student Responsible
Scientific-production enterprise "Vostok", Novosibirsk	1978	Layer-by-layer analysis of the impurity of different semiconductors	V.N. Polyakova
"Plant of pure metals", Svetlovodsk, Central Ukraine	1979	Purity of Ga: 10^{-6}–10^{-5}% of Sn, Ge; 10^{-7}% of Au	*V.E. Katuhin, N.T. Rud
Scientific Research Institute on Molecular Electronics, Moscow	1980	Determination of Zn, Cd, Pb, Cu, Bi, Sn, Te in films of AsGa; detn. of Hg in technological liquids	V.M. Pichugina, E.O. Portnyagina
Scientific Research Institute on Molecular Electronics, Moscow	1980	Determination of Zn, Cd, Pb, Cu, Bi, Sb, Sn, Te, Au, Ba in films of SiO_2, Si_3N_4 Al_2O_3 on AsGa base; layer-by-layer analysis of Se in AsGa	V.M. Pichugina, E.O. Portnyagina, N.M. Mordvinova, N.T. Rud
Scientific Research Institute "Pulsar", Moscow (the central institute for production of transistors in USSR)	1980	ASV* of different technological liquids and SiO_2	V.M. Pichugina
Siberian Physico-Technical Institute, Tomsk	1980	Concentration of Ga and As in films and powders of ZnS(GaAs), CdS(GaAs), AsS CdS(GaAs), ZnSe(GaAs)	N.T. Rud, N.M. Mordvinova
Scientific Research Institute of Semiconductor Devices, Tomsk	1980	New ASV* device called PP-1 was introduced in its laboratory	U.A. Ivanov, E.M. Kulagin

(*Continued*)

Table 5.4: (*Continued*)

Name of an Organization	Year	Type of Work	PhD Student Responsible
Scientific Research Institute "Pulsar", Moscow (the central institute on production of transistors in USSR)	1981	Determination of 10^{-5}–10^{-7}% of Ga, Ge,Te, As in technological solutions.	V.A. Kolpakov, N.T. Rud
Scientific Research Institute of Semiconductor Devices, Tomsk	1981	Development of the universal polarograph for determination of Cu, Sn, Cd and others in solutions and materials *used for production of semiconductor technology	E.M. Kulagin
All-Union Scientific Research Institute of Phosphor (referring to the light-issuing chemicals), Stavropol	1982	Determination of Cd, Pb, Sn, Cu, Sb in SnS and ZnSe	Z.S. Mihaylova, V.M. Pichugina
All-Union Scientific Research Institute of Phosphor (referring to the light-issuing chemicals), Stavropol	1983	Determination of Cd, Pb, Sn, Cu, Fe, Sb in powders of HfO_2, GeO_2, ZrO_2, TiO_2, GaF_3, BaF_2, Na_3AlF_6. Determination of V and Ag in BaF_2, CaF_2, Na_3AlF_6	N.T. Rud, N.M. Dubova, Z.S. Mikhaylova
Scientific Research Institute of Electronic Devices Production, Omsk	1983	Concentration of V in films of Al at the V base	N.M. Dubova
Scientific Research Institute of Electronic Devices Production, Omsk	1983	Concentration of V in films of SiO_2 at the V base	N.M. Dubova

(*Continued*)

(*Continued*)

No.	Name (Surname, Followed by First then Middle Names)	Year of Thesis Defence
33.	Sutyagina, Galina Nikolayevna	1973
34.	Mikheeva, Nina Petrovna	1973
35.	Chernova, Larisa Alekseyevna	1974
36.	Ivanov Igorevich, Yuri	1974
37.	Zheltonozhko, Anatoly Arkadievich	1974
38.	Dzhabarova, Naila Kashafovna	1974
39.	Veyts, Nellie Aleksandrovna	1974
40.	Gridaev, Viktor Fyodorovich	1974
41.	Gomza, Vladimir Vasilyevich	1974
42.	Chemeris, Marina Sergeyevna	1974
43.	Reznikova, Svetlana Stepanovna	1974
44.	Smolova, Larissa Mikhaylovna	1974
45.	Ryashentseva, Tatyana Fyodorovna	1975
46.	Belousov, Yuri Petrovich	1975
47.	Oblivantsev, Yuriy Nikolayevich	1975
48.	Baev, Vladimir Sergeevich	1977
49.	Blinnikova, Alexandra Alexandrovna	1977
50.	Podkorytova, Nellya Vasilyevna	1977
51.	Rahmonberdyev, Alisher Dzhamolovich	1978
52.	Geyneman, Arkady Edmundovich	1978
53.	Yargaeva, (Zgadova) Valentina Aleksandrovna	1979
54.	Chernyshova, Nina Nikolayevna	1979
55.	Prokudina, (Vishtakalyuk) Ludmila Dmitrievna	1979
56.	Dominova, Irina Gumerovna	1979
57.	Vakhmantseva, Nina Ivanovna	1979
58.	Zarubina, Raisa Fyodorovna	1979
59.	Mamontova, Inessa Petrovna	1979
60.	Kubrak, Boris Arkadievich	1980
61.	Pikula, Nina Pavlovna	1981
62.	Kolpakov, Valery Aleksandrovich	1981
63.	Rud, Nadejda Terentevna	1981
64.	Portnyagina, Ella Oskarovna	1982
65.	Botkina (Chubakova), Yelena Ivanovna	1982
66.	Mordvinova, Nina Mikhailovna	1982
67.	Mamayev, Anatoly Ivanovich	1983
68.	Dubova, Nadejda Mikhailovna	1984

(*Continued*)

(*Continued*)

No.	Name (Surname, Followed by First then Middle Names)	Year of Thesis Defence
69.	Bulantseva (Polyakova), Valentina Nikolayevna	1984
70.	Anisimova, Lyubov Sergeyevna	1984
71.	Zinovyev, Alexander Ivanovich	1984
72.	Kulagin, Evgeny Mikhaylovich	1984
73.	Mamaeva, Vera Alexandrovna	1985
74.	Shifris, Beye Solovna	1985
75.	Dmitrienko, Victor Petrovich	1986
76.	Mazur, Lyudmila Petrovna	1987
77.	Maksimova, Nina Innokentevna	1987
78.	Chertov, Sergey Vitalevich	1987
79.	Solovtseva (Shvets), Lyudmila Alexandrovna	1987
80.	Slepchenko, Galina Borisovna	1988
81.	Gindullina, Tatyana Mikhaylovna	1988
82.	Bondarenko (Arkhipova), Lyudmila Nikolaevna	1988
83.	Stas, Irina Evgenevna	1989
84.	Klyueva, Tatyana Borisovna	1989
85.	Obraztsov, Sergei Viktorovich	1989
86.	Hahanina, Tatiana Ivanovna	1989
87.	Sarajeva, Vera Emelyanovna	1990
88.	Svintsova, Ludmila Dmitrievna	1991
89.	Pustovoytova (Gubina), Nadejda Germanovna	1992
90.	Hustenko, Larisa Anatolievna	1993
91.	Svischenko, Natalya Mikhailovna	1994
92.	Vagina, Svetlana Eduardovn's	1994
93.	Ivanovskaya, Elena Alexandrovna	1994
94.	Abakshina, Svetlana Aleksandrovna	1994
95.	Korotkova, Yelena Ivanovna	1995
96.	Fofonova, Tatyana Mikhaylovna	1996
97.	Smirnova, Irina Vladimirovna	1996
98.	Chernova, Elena Evgenyevna	1997
99.	Romanenko, Sergei Vladimirovich	1998
100.	Leites, Elena Anatolevna	2000
101.	Bobleva, Yulia Vladimirovna	2002

6

Life in Tomsk After Retirement

Leaving the Department in 1985 to Retire

As the head of the department and the Special Research Laboratory, Stromberg spent most of his time on administration, planning, monitoring, reporting, etc. and increasingly less time on his scientific work. Stromberg began to think about taking a back seat role from these organizational duties, and to consider retirement. He finally decided to retire after the death of his colleague, Senior Lecturer Anna Abramovna Lelchuk. They had worked together for a long time: Anna had defended her thesis under the direction of Stromberg and they had written a book of problems in thermodynamics. In addition, she was a woman of his era: she had survived the war and the cult of Stalin. Her death prompted Stromberg to think about his own mortality: *'The shells are falling ever closer'* was a thought that was foremost in his mind.

Upon his retirement in 1985 Armin planned to appoint his most promising and energetic students (Ivanov and Kaplin) to management posts in the Department of Physical and Colloidal Chemistry, and of the Special Research Laboratory, respectively. The Department of Analytical Chemistry and Electrochemical Production was to be led by another of his former PhD students, Karbainov. However, tragic events threw Armin's carefully drawn-up plans into disarray. Unexpectedly, Ivanov died in 1986 of a severe form of cancer; then Kaplin died of a heart attack in 1989. Both the department and the laboratory were left without a leader. A new chairman of the department was appointed — Professor Dr Belikhmayer — but the themes of his scientific work (the kinetics and the mechanism of the low-temperature destruction of peat and other fuels) were completely different from his predecessors' scientific interests, where eight docents and Stromberg himself (as a professor-consultant) were specialists in the field of electrochemistry. In 1993, Belikhmayer emigrated to Germany, where he had relatives. After Belikhmayer's departure, the Department of Physical and Colloidal Chemistry was formally combined with the Department of the Technology of Silicates (headed by V.I. Vereshchagin), but it retained its previous scientific direction in the comprehensive development of anodic stripping voltammetric techniques.

After the untimely death of Kaplin, the Special Research Laboratory was divided into three parts. The first fell under the management of Karbainov (in the Department of Analytical Chemistry), while the other two sections joined the Department of Physical Chemistry in the form of Research Laboratory 504 led by Yu.A. Ivanov (working on the construction of voltammetric analysers), and Research Laboratory 506, which is currently (as of 2010) led by G.B. Slepchenko (concerned primarily with the stripping voltammetric analysis of foodstuffs and pharmaceuticals).

This period of uncertainty and change within the management structure of the department brought about by the deaths of several successful managers coincided with major political changes that were developing in the USSR at this time. By the end of the 1980s it had become apparent to those in the highest levels of the Soviet government that, unless serious political and economic change was brought about, the country was facing a serious catastrophe. So began Gorbachev's period of 'reconstruction' (perestroika).

Perestroika and Stromberg's Return to Nizhni Tagil

Gorbachev's policy of perestroika brought about great changes in the political, economic and cultural aspects of life for millions of ordinary Russians. Uncensored material and documents concerning the previous decades of state repression and the activities of the secret police (by then the NKVD had been superseded by the KGB) against ordinary Russian people began to gradually be released. A new era dawned on Stromberg. His daughter, Elza, recalls that:

> *'It was as if a veil fell from his eyes: the truth concerning his years of repression blinded him; the shadow of fear that he had been forced to live under for so long suddenly began to fall back. My father was gripped by a passionate desire to no longer "live a lie", but to learn the truth about his imprisonment, his demotion to a second-class Soviet citizen, and the reasoning behind it.'*

Thus, as the damn of forbidden literature burst open and the public, historians and journalists were granted access to the hitherto secret files and folders supplemented by uncensored materials, the first books and periodicals began to appear that described the nature of the Soviet system, the Great Terror and its victims, the Soviet Germans, and the working armies in which Stromberg had 'fought' during World War II. Stromberg sought to obtain and read all of these significant new books and periodicals.

Stromberg was also keen to re-establish contact with his former friends from the GULAG (up until this time it had been too dangerous for Armin to try to correspond with these former GULAG inmates even after the war had ended). He tried to contact Rauschenbach, Fridrikhsen, and Rikert (see Chapter 3). Then, in 1990, he learned about the opening of the first monument to the Soviet Germans of Special Force 18–74 who had been killed in Nizhni Tagil. Armin Stromberg was 80 years old at the time, but despite this he insisted on making the long return journey from Tomsk to Nizhni Tagil, along with his daughter, Elza, to honour the memory of the victims he had worked alongside during his imprisonment in Special Force 18–74 between 1942–1943. In Ekaterinburg, Stromberg's nephews — the scientists Andrey Ivliev and Mikhail Pokrovskiy — joined them for the journey northwards to the city of Nizhni Tagil. Together the group also managed to locate and re-establish contact with Fridrikhsen.

On the site of the NKVD GULAG complex of camps known as 'Tagillag' the group of Armin's surviving former 'working army soldiers' of Special Force 18–74 angrily recalled moments from their past lives, sketched a plan of their old barracks (shown in Chapter 3), and managed to find the ruins of the brick plant at which they had worked. An enormous

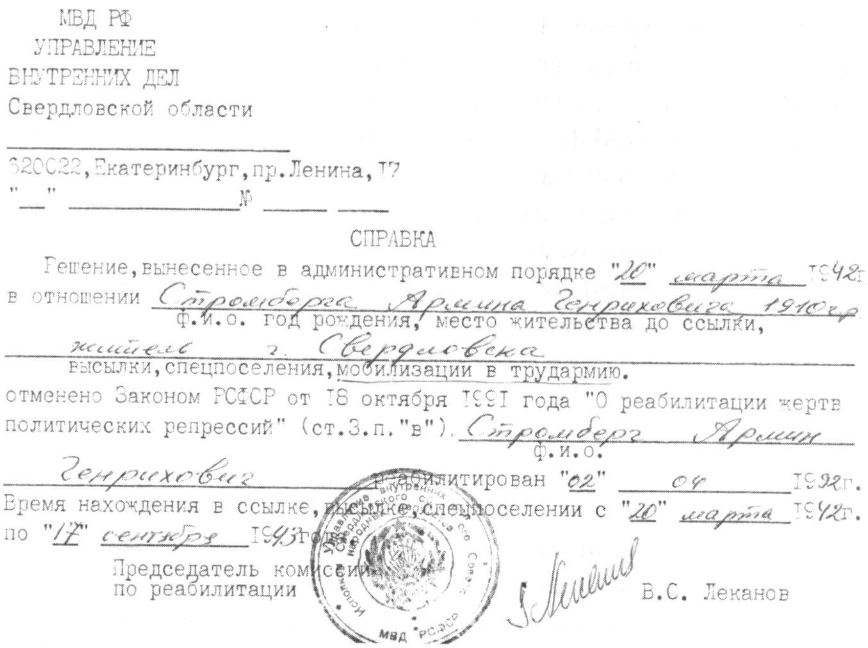

МВД РФ
УПРАВЛЕНИЕ
ВНУТРЕННИХ ДЕЛ
Свердловской области

620022,Екатеринбург,пр.Ленина,17
"__" ____________№ _____ ____

СПРАВКА

Решение,вынесенное в административном порядке "20" марта 1942г.
в отношении Стромберга Армина Генриховича 1910г.р.
ф.и.о. год рождения, место жительства до ссылки,
житель г. Свердловска
высылки,спецпоселения,мобилизации в трудармию.
отменено Законом РСФСР от 18 октября 1991 года "О реабилитации жертв политических репрессий" (ст.3.п."в"). Стромберг Армин
ф.и.о.
Генрихович реабилитирован "02" 04 1992г.
Время нахождения в ссылке,высылке,спецпоселении с "20" марта 1942г.
по "17" сентября 1943года.

Председатель комиссии
по реабилитации В.С. Леканов

The certificate of formal ‘rehabilitation’ received by Armin Stromberg in 1992, reinstating him to full citizenship.

rights within these states, including Russia. It was as a result of these cathartic exercises, and of meeting former German prisoners of war, that Stromberg began to turn his analytical mind to studying the repression of his fellow Russians in a rigorous, scientific manner. It was a task that would occupy him for much of his retirement.

With perestroika came a certain amount of political freedom to the previously repressed Russian people, but it also created economic chaos. The legalization of private ownership of property and business was introduced to move Russia from the previous Stalinist system over to some form of a private market economy. However, this led to periods of hyper-inflation in the early 1990s. Pensioners, such as Stromberg, saw their savings, carefully accrued over many years, wiped out by repeated devaluations of the rouble and massive price inflation of goods and basic services. Armin did not own any property — all he had were his savings that were now almost worthless! Yet the events of Armin's life had taught him not to value material possessions or money. As a young boy he had watched his mother lose everything that the family had ever owned in the Civil War that followed the October Revolution, and even this was nothing to her compared with the loss of her husband. So, Stromberg's diaries and records from these years simply note that he found it necessary to stock up on enough provisions as he could find, to economize these as best as he was able, and, perhaps Armin's greatest fear after his time spent in the NKVD camp, to await the onset of hunger.

Science in Retirement: Teaching an Old Dog New Tricks

The economic and political turbulence that came hand in hand with the introduction of perestroika also led to a serious decline in the funding and development of Russian science, culture, and public institutions such as libraries, museums and universities. The funds allocated to universities during the 1990s were paltry. As academic scientific research is heavily reliant on good access to the literature in the form of specialist periodicals and journals, it was a matter of great concern to many Russian scientists during this period that libraries and repositories were no longer able to afford subscription to many of the principle scientific journals. In TPU, at this time, even the basic flow of information usually obtainable via commonly used literature database services such as the Russian Abstracts of the Journal of Chemistry was frequently interrupted! One must also consider that these were paper archives, whose physical storage and delivery is more expensive than the modern electronic equivalents. The widespread use of desktop computers and the Internet, tools that allow modern scientists access to electronic literature databases and search engines (and that we take for granted today), only began to appear in Russia at the very end of the 1990s — much later than in other Western countries, than in Europe and the US, for example.

As is usually the case it was scientific research into fundamental problems of interest (as opposed to applied research to solve a specific problem or target industrial applications) that suffered most of all. Yet it is largely from 'blue sky' fundamental research that new applications and scientific discoveries are made — a lesson that some Western governments, reeling from the economic crises of 2008–2010, seem not to have learned. The Russian government continued to subsidize the development of applied science in scientific research institutes, but the Russian universities were supported only by grants, and these were almost negligible compared to the amounts the universities needed to cover their basic running costs, let alone fund anything other than the most basic research. Added to this was the Moscow-centric attitude of the funding bodies that saw the vast majority of any available funds allocated to the Muscovite universities and institutions. This further aggravated the decline in science within the more 'provincial' universities that served the vast remainder of Russia outside of Moscow.

The plight of Russian science and education was further complicated by an unusual demographical problem. As a result of the revolution of 1917, the process of cultural heritage, where shared knowledge and intellectual ideals are passed from one generation to another, was interrupted. The direct descendants of anyone educated in the pre-revolutionary Russian culture had died or were of advanced age by 1990. The old 'educated classes' had been replaced by a 'Soviet intelligentsia', that was seen by many both inside and outside of the former USSR as being somewhat defective in comparison. The Soviet intelligentsia was comprised only of individuals educated 'the Soviet way', following Soviet ideals and brainwashed to be naturally xenophobic, suspicious or dismissive of any ideas originating from outside the Soviet sphere of influence. This had a profound influence on the process of education in schools and institutes, which had become somewhat stifled and repetitive. There was a 'brain-drain' of gifted young people moving away from academia. Many young people, whom upon leaving the Soviet education system still retained some semblance of initiative-taking and entrepreneurial skill, were attracted into the newly created spheres of semi-legal private business and commerce — though this was an area rife with extortion and criminality. Those school and university leavers whom retained an interest in a career in science increasingly tended to search for positions abroad, since it was almost impossible to live on the beggarly wage of a Russian scientist, and worse still, impossible to conduct serious studies in the absence of proper funding for equipment.

As a professor-consultant (emeritus-professor) Stromberg certainly did not cease his scientific research activities, but his focus became slightly different. His former PhD students became doctors of science and led the work in their own groups. So, Armin began to work mainly in the library, either alone or with one or two colleagues from the department. The first task that he set himself was to collate a database of several thousand abstracts covering all trends in the development of stripping voltammetry over the period from 1985 onwards, and to upload this database into the memory of the still imperfect Soviet computer 'Iskra-30'. It was his intention that the assembled information could to be sold to other scientific institutes or universities. Unfortunately, the Iskra-30 computer was prone to crashing and breaking down, and Armin's pet project was superseded by the subsequent introduction of the Internet.

A similar project, undertaken by Armin between 1989–1994 together with the docent and former student of Stromberg's, Boris Nazarov, and a TPU librarian, was to assemble a bibliographical directory on the theory of stripping voltammetry. One should note that at this time there were no computers to allow one to search for thousands of articles in a matter of seconds. The abstracts of the thousands of papers published had to be painstakingly searched by hand, analysed by the electrochemical methods used, experimental conditions, mechanistic theory developed, results and conclusions, etc., and the bibliography written out on a typewriter in the library. 'The list' was published through the TPU publishing house. Stromberg and Nazarov had intended to prepare a large survey on the theory of stripping voltammetry, yet, despite sponsorship of the project by a small Tomsk firm (Tekhnoanalyt), insufficient funds were raised and the full survey was never written.

Stromberg worked on the final and most interesting project of this period (from 1995 up until his death in 2004), marking a departure from his familiar territory of developing analytical solutions to the theory of stripping voltammetry into the modern realms of computer-aided numerical analysis. This occurred as a result of the number of changes within the department. First, a graduate of Tomsk State University, Sergey V. Romanenko, was accepted into the department as a postgraduate student. Although Romanenko was formally supposed to be specializing in 'analytical chemistry', he was interested in, successfully passed, and later

Stromberg pictured with his students Serge and Ella Romanenko discussing a scientific paper.

taught specialist courses in 'information theory', where a profound knowledge of mathematics and computer technology was imparted to the students.

As part of his management of the graduate school in the department, Armin became interested in Romanenko's work and later they were able to attract a number of capable students who shared an interest in applying computational methods to solve complex analytical chemical problems. This partnership allowed Armin to return to some earlier studies concerning the theory of analytical signals in stripping voltammetric analysis. Armin had begun to develop theory on this matter, and had attempted to solve the subsequent equations analytically with two graduate students in 1974 and again in 1981 — long before the availability of personal computers in the USSR. However, by 1995, the use of contemporary PCs made it possible to considerably increase the volume of calculations and to raise the simulation of analytical signals to new qualitative and quantitative heights. Thus, computer-aided numerical simulations (approximate solutions to equations that are analytically insoluble, i.e. they have no known exact mathematical solutions) allowed the team to rapidly analyse the shapes and features of current–voltage curves containing more than 100 points at intervals of 2 mV. Using this new (for the Tomsk school at any rate) approach, Romanenko was able to successfully defend his thesis and obtain his doctorate degree.

By the end of the 1990s, Stromberg himself had also acquired a personal computer and, whilst still in his mid-80s, had mastered some of its operations and taught himself rudimentary programming! Who says that you can't teach an old dog new tricks? This new field of computer simulation inspired Stromberg and Romanenko to explore several new ideas, to attract a number of masters and postgraduate degree students to work on them. The pace and intensity of Stromberg's scientific output grew anew, with each year between 1997–2004 seeing increasing numbers of articles published in the scientific journals. It was a source of great happiness to Stromberg when his graduate student, Romanenko, published his first paper with Armin in a foreign scientific journal.[1]

Meanwhile, the intensity and manner in which scientific work was carried out under the new economic conditions began to change in what had once been known as the Special Research Laboratory. The research carried out within the Analytical Chemistry Department and in the research laboratories that had once formed Stromberg's Special Research Laboratory necessarily became ever more applied in nature. The interlaboratory relationships within the department also began to alter. Instead of the laboratories operating as large groups of collaborative researchers as they had done in the past, they split into small groups of researchers working in isolation from one another. The attendance rates at scientific seminars began to fall off; glasnost, or openness in the scientific sense, between the various groups and laboratories disappeared. Due to the material lack of funds in the research organizations many scientists did not have the opportunity to visit national or international conferences leading to a degree of scientific isolation. Traditional conferences on electrochemical topics were no longer held in Tomsk. All of these changes

---

[1] A.G. Stromberg, S.V. Romanenko, 'Determination of the true form of overlapping peaks, deformed by the base line in the case of stripping voltammetry', *Fres. J. Anal. Chem.*, 1998, 361, 276–279.

were very disturbing to Stromberg, and prompted him to act. He criticized the formalism of reports written by graduate students, and he re-wrote 'the rules' on how their scientific research should be presented and discussed. Stromberg himself began to lead the scientific seminars and actively participated in the discussion of scientific matters.

Curiously, the new head of the united Department of Analytical and Physical Chemistry at TPU was technically an organic chemist. Despite this apparent shift in the direction of the laboratory's research areas, Stromberg strongly supported the departmental head's initiatives when he began to intensively equip the department with computers, organize computer classes with the help of Romanenko's group and roll out a programme to increase the number of graduate students in the department. He also managed to reinstate the organization of regular all-Russian and international electrochemical conferences held at TPU in Tomsk.

Still, the style of management concerning graduate students and the principles of the former Special Research Laboratory had completely changed. The organization and principles upon which Stromberg had founded the School of Electroanalysis in Tomsk, and the Special Research Laboratory in particular, in 1962 were unrecognizable in the department of the 1990s. Stromberg found these changes painful to observe, but reflected on them somewhat philosophically. He considered that:

> *'Each organization is subjected to the same stages of life and evolves as a living organism does: inception, youth and growth, ageing and decline, and finally death through the disintegration of the organization as the basic direction, energy and driving forces used to create the organization in the first place are exhausted.'*

The years between 1998 and 2000 were hard for Armin. His health was deteriorating and he found that he could only rely on himself to develop his interests in his own scientific direction. Despite these hardships, Stromberg still had to play the role of the chief-organizer and head of the 'Symposia on the Theory and Practice of Stripping Voltammetry', including undertaking many arduous tasks associated with the celebrations of the hundredth anniversary of the founding of Tomsk Polytechnic University in 2000.

Hobbies

Aside from his scientific activities, music played a large part throughout Stromberg's life. For example, he studied the piano in his childhood with private teachers, played the violin sonatas of Beethoven and Grieg with his stepfather, and later enjoyed playing arrangements of many symphonies and operas as piano duets with his daughter, Elza. His favourite composers were Chopin, Beethoven and Grieg. There had always been a piano in his home, one that Stromberg had rented for years. When he was 84 he finally purchased his own piano and began to play almost every day, learning 13 preludes by Chopin off by heart over the next 10 years.

Stromberg kept himself physically active in his later years, for example by clearing snow from the pavements around his apartment in winter, and gardening in the summer. Stromberg found that tearing out weeds and pruning back trees was a good way for him to relax, as well as having the benefit of a tidy yard. As always, these were meticulously pre-planned activities scheduled to last for one and a half to two hours each and every day.

Finally, Stromberg's interminable love was the library. Every week Armin rushed to browse through the exhibition of new literature added to the library's collection. His interests were rather broad: aside from books on chemistry he was drawn to books on history, philosophy and culture. Armin would photocopy the title and table of contents of any book that he was interested in. Thus, he amassed a collection of references of more than 800 titles, which he catalogued by theme. In doing this he was able to combine all the major joys of his life: books, classification and the discovery of new things!

Vita Brevis Est… Ut Volito Oportet![2]

As part of the reconstructions and reforms occurring in many spheres of Russian life during the 1990s, there arose a renewed interest in recognizing the victims of the tragic history of the former totalitarian Soviet state. Armin was frequently interviewed about the events of his life in the media

[2]Roughly translated as: 'Life is short … it is necessary to hurry!'

Armin Stromberg at the opening of the garden courtyard and alley bearing his name in 2000.

and for contemporary Russian history books. Consequently, Stromberg was the recipient of more awards, honours and titles over a period of seven years in the 1990s, after his retirement, than he had received during his entire working life spanning the previous seven decades.

He was presented with a medal in recognition of '*his valiant labour during the Great Patriotic War*' in 1993 at the age of 83. In recognition of his enormous contribution to science, particularly in Tomsk, he was awarded, aged 84, the Honourable Title '*the Deserved Soros' Professor*' in 1994, and in 1998 the Honourable Title '*the Honoured Professor of TPU*'. Stromberg's civic duties, in particular his voluntary tending of the communal gardens surrounding his apartment block (for which he won awards and plaudits from the Mayor of Tomsk), were also recognized by his colleagues, students and former graduates of the Special Research Laboratory in celebration of his ninetieth birthday in 2000. A small garden courtyard was established in front of the Physical and Analytical Chemistry Department at TPU and a path leading from this garden was named 'Stromberg Alley' in his honour. Whilst Stromberg was evidently very touched by these accolades, particularly the garden courtyard, he wryly noted in private that:

> *'Imagine that I had not lived for 23 years beyond the average age of men in the Russian Federation, and instead had died at the respectable age of 80. I would not then have learned that I was worthy of receiving so many honourable and distinguished titles.'*

When Armin's colleagues and students had gathered for the opening of 'Stromberg Alley', Armin suggested to them that he would like to mark the fortieth anniversary of the founding of the Special Research Laboratory in 2002 by writing a book called *Who's Who in the Special Laboratory of Microcontaminant Detection.* So a committee was organized to work on questions for a survey that was then sent to 90 addresses of former members of the Special Research Laboratory. The collection of this material was long and difficult; reminders had to be sent out with monotonous regularity. Eventually, the bulk of the material had been collected, but it was arranged in a rather fragmented fashion.

Despite his failing health, and true to form, Armin set about putting the data into some semblance of order and wrote it up himself in order to give it a single, uniform style. He wrote about his own life and his recollections of his first postgraduate student in Tomsk, Vitya Igolskij. Armin became completely absorbed by this task. He felt that he had little strength left, and wanted to create a sum total of his life, to bring all of his students 'under one roof'. He hurried to get the book finished — quite often his relatives would force him to spend two or three hours a night working at his computer — as one biography after another was compiled and edited. The history of the creation of the Special Research Laboratory was carefully thought over and printed, as were its principles of work and structure, its advantages and disadvantages. Such were Armin's principles of work and life as a scientific leader. The results of the Special Research Laboratory's scientific activities were summarized, and annotations of 100 dissertations on the development of the method of stripping voltammetry were collected. The recollections of scientists on their lives and work and about the united scientific front that embodied the ethos of the Special Research Laboratory were collected. Thus, after a collective effort, a book of almost

300 pages began to appear: *'The Fortieth Anniversary of the Tomsk Electrochemical School and the Special Laboratory of Microcontaminants Detection. The Development of Stripping Voltammetry. The Preparation of More Than 100 Candidates of Chemical Science in the Special Laboratory of Microcontaminant Detection.'* And yet it was still not completely finished by the time of the anniversary in 2002, and would take a further two years to be published.

By 2004, even though his health was deteriorating and he was becoming increasingly weak, Armin was full of plans and was busily planning studies and reading and writing dozens of articles. During the final years of his life, he supervised his last student, Ekaterina Larionova, through the preparation of her PhD thesis. Sadly, she would have to defend it without him. He worried that she would not be able to find a job afterwards, and wrote a report to the dean about this. He also discussed his colleagues' scientometric data with interest. For example, I.V. Zibareva of the Siberian branch of the Russian Academy of Science had dedicated her work to studying the history of the development of voltammetry and stripping voltammetry both in Russia and worldwide. Her conclusion at the time was that, when considering the number of publications concerning stripping voltammetric analysis, Armin Stromberg was ranked third in the world. She claimed that only Professor Joseph Wang (USA) and Professor Richard G. Compton (Oxford) had published more articles on the subject.[3] Meanwhile, Stromberg was impatiently awaiting the publication of his final book concerning the history of the Special Research Laboratory.

The collection and printing of this compilation took until September 2004. Finally, the advance copy was carefully examined, but the entire press run was yet to be printed and Armin was worried. He wanted to get all his former students and colleagues together on his ninety-fourth birthday on the 16th of September 2004 to personally present them with a copy of the unique history of their Special Research Laboratory, in which each and every one of his PhD students was mentioned. However, his health was deteriorating with each passing day. Finally, on the 15th of September, the finished press run was delivered to his home. The people closest to

[3]The authors would like to stress here that this was the conclusion of Zibareva's own analysis and may not be factually accurate.

him knew that Armin did not have the strength to sign all 100 copies and so photocopies of his signature were carefully stuck into each copy of the book.

At 4.00 p.m. on the 16th of September, the day of Armin's ninety-fourth birthday, around 40 people, including his postgraduate students and colleagues, gathered in the dining room at Armin's house. Armin, who was very weak by then, was brought in on a wheeled computer chair and it was clear that he would only be able to manage to remain amongst the party for half an hour without needing a rest. A doctor, Raisa Afonina, was waiting nearby with syringes, ready to help should the need arise. With difficulty, Armin said his words of thanks and expressed his wishes and gratitude. Then he presented everyone there with a copy of his last book, the sum total of his life. Everyone drank champagne and made toasts to his health. They wished him happy birthday and congratulated him on the publication of his book. He was proud and happy. He had done everything that he could with regard to this last book, and he had been waiting to be able to give it to his students. He wanted them to work as hard and as purposefully following their graduation as they had in the laboratory and to remember their youth and their alma mater, the Special Research Laboratory. All those present recognized that this was a uniquely special moment.

Armin passed away two days later. He had lived until the age of 94 thanks to his great willpower that ordered his body to carry on and to finish his life's work. In life he had demonstrated exemplary spirit, perseverance and persistence in overcoming the many difficulties and hardships he and his loved ones had faced. Through his science, Stromberg has made some of the most important and prolific contributions to the field of analytical chemistry and stripping voltammetry, although this is still largely unknown in the West. He trained generations of Russian science students, many of whom went on to become successful and noteworthy chemists in their own right. He may have lived his life as a second-class citizen, and become a first-class scientist, but he was above and beyond all a first-rate man.

